T0271001

Coating Application for Piping, Valves and Actuators in Offshore Oil and Gas Industry

This book looks at the applications of coating in piping, valves and actuators in the offshore oil and gas industry. Providing a key guide for professionals and students alike, it highlights specific coating standards within the industry, including ISO, NORSOK, SSPC and NACE.

In the corrosive environment of a seawater setting, coatings to protect pipes, valves and actuators are essential. This book provides both the theory behind these coatings and practical applications, including case studies from multinational companies. It covers different offshore zones and their corrosivity level alongside the different types of external corrosion, such as stress cracking and hydrogen-induced stress cracking. The key coatings discussed are zinc-rich coatings, thermal spray zinc or aluminum, phenolic epoxy and passive fire protection, with a review of their defects and potential failures. The book also details the role of coating inspectors and explains how to diagnose faults. Case studies from companies such as Aker Solutions, Baker Hughes, Equinor and British Petroleum illustrate the wide range of industrial applications of coating technologies.

This book is of interest to engineers and students in materials, coating, mechanical, piping or petroleum engineering.

Coating Application for Piping, Valves and Actuators in Offshore Oil and Gas Industry

Karan Sotoodeh

CRC Press
Taylor & Francis Group
Boca Raton London New York

CRC Press is an imprint of the
Taylor & Francis Group, an **informa** business

First edition published 2023
by CRC Press
6000 Broken Sound Parkway NW, Suite 300, Boca Raton, FL 33487-2742

and by CRC Press
4 Park Square, Milton Park, Abingdon, Oxon, OX14 4RN

© 2023 Karan Sotoodeh

CRC Press is an imprint of Taylor & Francis Group, LLC

ISBN: 978-1-032-18719-8 (hbk)
ISBN: 978-1-032-18722-8 (pbk)
ISBN: 978-1-003-25591-8 (ebk)

DOI: 10.1201/9781003255918

Typeset in Times
by codeMantra

Contents

Preface

The book is about applications of coating for the piping, valves and actuators in the offshore oil and gas industry. The offshore or marine environment is very important since two-thirds of the earth's surface is made of water. The sea has been a source of food and energy for humans for millennia. Today, great attention is paid to the marine environment from an energy point of view, as many oil and gas resources are located under the seabed. Offshore environment is extremely corrosive due to seawater and existence of chloride in the atmosphere. One of the most important ways to protect offshore facilities such as piping, valves and actuators is by applying coating. The book is in six chapters. Chapter 1 covers different offshore zones and their corrosivity level as well as the basic concept of corrosion. In addition, different types of external corrosion types such as general, localized (crevice, pitting, galvanic, corrosion under insulation) and environment cracking corrosion such as stress cracking corrosion and hydrogen-induced stress cracking are covered in this chapter. Chapter 2 addresses different techniques of surface preparation before applying the coating such as acid or chemical cleaning, water jet cleaning, sandblasting, etc. In fact, proper surface cleaning of the metal before applying coating is the most important step in coating applications. In addition, different surface contaminations such as oil, grease, rust and mill scale are covered in this chapter. Chapter 3 discusses briefly the coating standards such as ISO, NORSOK, SSPC, NACE, etc. The most important coating systems which are used for piping, valves and actuators are discussed in this chapter. These coatings are zinc-rich coating, thermal-spray zinc or aluminum, phenolic epoxy, passive fire protection, two-component epoxy for subsea applications, etc. Chapter 4 reviews coating defects and failures as well as the essence of coating inspection. The role of coating inspector and the required skills and knowledge for a coating inspector are discussed in this chapter. Chapter 5 provides a comprehensive review of the valve and actuator technology in the offshore oil and gas industry. The most important types and valves are reviewed in this chapter. Chapter 6 includes all the lessons learned and experiences of coating systems on the valves, actuators and piping in Norwegian and international offshore projects. Different experiences such as coating failures experience as well as solutions to prevent the failure, extension of coating on components, etc. are included in this chapter.

Author

Karan Sotoodeh, PhD, worked for Baker Hughes at his last position as a senior/ lead valve and actuator engineer in subsea oil and gas industry. He earned a PhD in safety and reliability in mechanical engineering at the University of Stavanger in 2021. He has almost 16 years of experience in the oil and gas industry, mainly with valves, piping, actuators and material engineering. He has written four books about piping, valves and actuators and more than 30 papers in peer-reviewed journals. He has been also selected at international conferences in the United States, Germany and China to discuss valves, actuators and piping. He has worked with many valve suppliers in Europe in the United Kingdom, Italy, France, Germany and Norway. He loves traveling, running and spending time in the nature.

1 Types of Corrosion in the Offshore Environment

1.1 INTRODUCTION TO THE MARINE ENVIRONMENT

The offshore or marine environment is very important since two-thirds of the earth's surface is made of water. The sea has been a source of food and energy for humans for millennia. Today, great attention is paid to the marine environment from an energy point of view, as many oil and gas resources are located under the seabed. The main characteristic of seawater is that it contains salt and chloride. In fact, the salinity of seawater comes from chloride, and the corrosivity of seawater is also largely attributed to chloride, as will be explained later in this chapter. Dissolved gases such as oxygen also play important roles in the corrosivity of seawater. In fact, the presence of oxygen in seawater increases the seawater corrosion rate, while oxygen-free seawater is not corrosive at all. The concentration of dissolved oxygen in the water is affected by multiple factors, such as temperature, amount of seawater movement, seawater contact with air and human activities. The amount of oxygen concentration increases in seawater with decreasing temperature. In other words, cold water can have more dissolved oxygen compared to warm water. Oxygen has been declining in the marine environment since the 1960s because of human activities, such as burning fossil fuels and discharging agricultural and human waste into the earth's oceans and waterways. Decreasing oxygen in the oceans is detrimental to marine life. The threat of oxygen reduction in the oceans from industrial activities has been known for many years.

Other corrosive compounds that may exist in seawater include carbon dioxide (CO_2) and hydrogen sulfide (H_2S). Seawater often contains hydrogen sulfide produced by sulfate-reducing bacteria. Both CO_2 and H_2S are acidic gases that reduce the pH level in seawater by reacting with it and forming carbonic and sulfuric acid, respectively. The corrosivity of the marine environment is not limited to the seawater itself. The air above the sea contains chloride taken from the sea spray. Thus, the marine environment is much more aggressive and corrosive compared to the land-based environment. However, the key point is that land-based areas close to the sea, i.e., coastal areas, are corrosive environments too.

The extreme corrosivity of the offshore environment necessitates the usage of coating for the protection of offshore facilities and components, such as valves, piping, pressure vessels, actuators and steel structures. Valves are mechanical components installed in piping systems to start or stop the fluid, control and

DOI: 10.1201/9781003255918-1

regulate the fluid, e.g., by preventing it from flowing back to an upstream section, and for safety purposes. Actuators are mechanical or electrical devices that move the valves. Offshore maintenance is very difficult, and its cost is extraordinarily high. Thus, it is important to select high-quality coating and ensure that it is applied correctly and under the right conditions.

When it comes to the subsea oil and gas industry, its components and facilities are completely submerged in seawater, which is one of the most severe and corrosive environments on the planet. In the subsea sector, the external protection provided by coating is combined with cathodic protection. The main objectives of coating selection in the offshore industry are as follows: compliance with health, safety, environment (HSE) and governing requirements; optimal protection of the installed components; minimal need for maintenance; environmentally friendly coating selection; and finally standardization and simplification. It should be noted that coating is used to protect facilities and structures as well as components against *external* corrosion. Thus, *internal* corrosion from corrosive and undesirable compounds, such as carbon dioxide and hydrogen sulfide, cannot be mitigated by the application of coating. Figure 1.1 illustrates an offshore platform exposed to seawater splash and the corrosive offshore atmosphere.

1.2 OFFSHORE ZONES

The intention of this section is to introduce the different offshore zones and their corrosivity values. There are four zones in the offshore industry, as illustrated in Figure 1.2: (1) atmospheric or topside, (2) splash, (3) tidal and (4) seawater or

FIGURE 1.1 Offshore platform and splash of seawater. (Courtesy: Shutterstock.)

FIGURE 1.2   Offshore zones. (Courtesy: Shutterstock.)

immersion. Structures can be located in any of these four zones. However, valves, piping and actuators are typically installed either in zone 1, the topside or atmospheric zone, or zone 4, the seawater or immersion zone.

In the *atmospheric zone*, the corrosion rate on unprotected steel is typically in the range of 80–200 μm (3–8 mils) or (0.076–0.20 mm) per year. The corrosion in the atmospheric zone is due to the presence of chloride and moisture in the environment. The other corrosion contribution factor is the effect of ultraviolet (UV) light. There are two degradation effects associated with UV light: one is the weight loss of steel, such as carbon steel, under UV light from the sun, which can be proven through electrochemical measurement; the second type of degradation is related to mechanical property loss when the coating polymer breaks down, rendering the coating useless.

In the *splash zone*, the structure alternates frequently between being above and below the waterline due to waves and the tide. The corrosion rate in this area is higher than in the atmospheric zone. The estimated corrosion rate in the splash zone is 200–500 μm (8–20 mils) or (0.20–0.50 mm) per year. Factors that exacerbate the corrosion rate in the splash zone are a mixture of corrosion and erosion from the seawater, UV light, debris and ice in some cases.

The *tidal zone*, like the splash zone, alternates between being above and below water frequently and is subject to the same causes of corrosion as the splash zone. The tidal zone is divided into two zones: high tidal and low tidal. The corrosion rate in the high tidal zone is higher than in the low tidal zone. The corrosion rate in the high tidal zone is more similar to the corrosion rate in the splash zone. The corrosion rate in the low tidal zone is lower and could be closer to that of the seawater or immersion zone.

In the *seawater or immersion zone*, which is always underwater, the corrosion rate is close to that of the atmospheric zone and lower than that of the splash zone. The corrosion rate is typically 100–200 μm (4–8 mils) or (0.1–0.2 mm) per year. Seawater is highly corrosive to mild steel, as it contains salt (e.g., 3.4%) and chloride. The reason why the corrosion rate is lower in seawater compared to the splash zone could be due to the former's lower concentration of oxygen. The corrosivity of seawater depends on different parameters, such as chloride or salinity level, oxygen concentration, temperature, the presence of biological organism, etc. In fact, increasing the oxygen, chloride level and temperature increases the seawater corrosion rate. Coating protection in the seawater or immersion zone can be combined with cathodic protection. The other important factor that should be taken into account in seawater is fouling. Fouling refers to the growth or formation of deposits or microorganisms such as bacteria on the surface of subsea structures and components. Fouling can be categorized into four types: (1) crystallization fouling due to the deposition of calcium carbonate or other types of salts, (2) corrosion fouling related to the oxidation or corrosion product, (3) biological fouling due to bacteria or even macroorganisms such as mussels and (4) particulate fouling due to different types of silt, mud and sand in the seawater.

The important point about the corrosion rates provided for the four zones is that the values only address general corrosion rates. Seawater and the marine environment can cause other types of localized corrosion types that are discussed in Chapter 1, such as crevice and pitting corrosion, which are not taken into account in the given corrosion rate values. Table 1.1 summarizes the corrosion rates in all four offshore zones and provides the final corrosion rates during the life of the plant, assuming a 20- or 25-year plant design life.

TABLE 1.1
Corrosion Rates (Metal Loss) for Offshore Zones

Offshore Zone	Corrosion Rate per Year (μm, mils, mm)	Corrosion Rate over 20 Years (mm)	Corrosion Rate over 25 Years (mm)
Atmospheric zone	80–200 μm (3–8 mils) (0.076–0.20 mm)	1.52–4	1.9–5
Splash zone	200–500 μm (8–20 mils) (0.20–0.50 mm)	4–10	5–12.5
Tidal zone	200–500 μm (8–20 mils) (0.20–0.50 mm)	4–10	5–12.5
Immersion zone	100–200 μm (4–8 mils) (0.1–0.2 mm)	2–4	2.5–5

1.3   BASICS OF CORROSION

The main objective of this section is to explain and describe the basics of corrosion. Humans have struggled constantly with the forces of nature for time immemorial, and corrosion is one such force. Corrosion is defined as the degradation of a material as a result of its interaction with the surrounding environment. Although this definition is applicable to any type of material, it is typically used for metallic alloys. Corrosion is an undesirable phenomenon with many negative consequences, including loss of human lives, loss of asset, loss of production and efficiency and environmental pollution. According to a National Association of Corrosion Engineers (NACE) estimate released in 2016, corrosion costs 2.5 trillion USD in 2013, which is equivalent to 3.4% of the global gross domestic product (GDP). An example of human death due to corrosion occurred in 1967 in West Virginia when a bridge collapsed, killing 46 people. A type of corrosion called stress cracking corrosion (SCC) caused the accident; SCC is explained in more detail later in this chapter.

Corrosion occurs because of the formation of electrochemical cells. An electrochemical cell is defined as a device capable of either deriving electrical energy from chemical reactions or facilitating a chemical reaction by introducing electrical energy. A common type of electrochemical cell is a battery, also known as a galvanic cell, in which electrical electricity is produced from a chemical reaction. An electrochemical corrosion cell includes four necessary factors. In other words, corrosion requires four elements to occur. These are listed below; if one of the elements or factors is eliminated, corrosion will not occur.

1. *An anode* is a metal from which an electron is emitted and metal loss occurs. An anode is less noble and has more negative potential compared to a cathode. Corrosion and loss of an electron occur in an anode in favor of the cathode if the two are in contact. Since an anode discharges electron, it becomes positive in the electrochemical cell. Oxidation, meaning the process of losing electrons, occurs in the anode.
2. *A cathode* is a metal in which a gain of electrons occurs. The electrons lost by the anode are received by the cathode. Therefore, the cathode is protected from corrosion thanks to the sacrificial metal in the anode. The cathode is negative in an electrochemical cell because it receives the electrons. Reduction is a chemical reaction that takes place in the cathode, involving the loss of oxygen atoms and the gain of one or more electrons.
3. *An electrolyte* is a substance that produces electrical continuity and conductivity in a solvent like water, and that plays a vital role in the occurrence of corrosion. When an anode and cathode are immersed in an electrolyte, the reaction between the two metals is triggered and accelerated. Some of the main electrolytes are sodium, chloride, potassium, calcium and magnesium. When these electrolytes are dissolved in water, they are separated into positive and negative ions called cations (+) and

anions (–), respectively. The anions are drawn to the electrode with a deficit of electrons, and the cations move in the opposite direction. In other words, the cations are positive and move to the negative terminal (cathode), and the anions are negative and move to the positive terminal (anode).

4. *A metallic path* must be in place between the anode and cathode. In fact, the anode and cathode should be electrically connected. The metallic path transfers the flow of electrical current from the anode to the cathode. The amount of current flow in the metallic path is proportional to the amount of metal loss or corrosion in the anode. The electron flow from the anode to the cathode is due to their difference in electrical potential. Figure 1.3 illustrates a corrosion cell that includes all four essential components. It is important to know that the electrochemical cell and the movement of the electrons are driven by the electrical potential difference between the cathode and the anode.

To summarize the theory behind an electrochemical corrosion cell, an electrochemical cell reaction is produced when an oxidation-reduction (redox) chemical reaction occurs in which electrons are transferred between the anode and cathode through a metallic path, which could be a wire. Because the reactions are physically separated from each other, each individual part, meaning anode or cathode, is called a *half-cell*, and each reaction is called a *half-cell reaction*. Equations 1.1

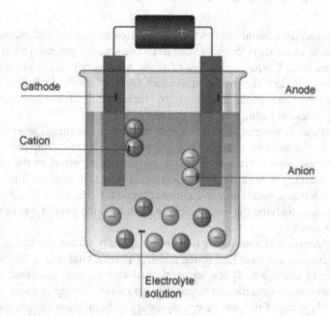

FIGURE 1.3 Electrochemical corrosion cell including all four essential elements. (Courtesy: Shutterstock.)

and 1.2 show how oxidation takes place at the anode and reduction occurs in the cathode.

Oxidation at the anode

$$\text{Reduced element or species} \rightarrow \text{Oxidized element or species} + ne^- \qquad (1.1)$$

Reduction at the cathode

$$\text{Oxidized element or species} + ne^- \rightarrow \text{Reduced element or species} \qquad (1.2)$$

The corrosion cell potential, known as E_{Cell}, is calculated according to Equation 1.3 and depends on the electrical potential of the anode and cathode.

Electrochemical corrosion cell electrical potential (E_{Cell}) calculation

$$E_{Cell} = E_{Cathode} - E_{Anode} \qquad (1.3)$$

The electrical potential of the anode and cathode are measured in practice with respect to the standard hydrogen electron with zero-volt electrical potential. Thus, it is essential to measure the standard electrical potential of each element with hydrogen to provide a standard framework. The next important point is related to the conditions in which the measurement of electrical potential takes place. This condition is defined as 1 molar solution on both anode and cathode, a gas pressure of 1 atmosphere (101.325 kPa or 14.5 psi) and a temperature of 25°C. Figure 1.4 provides examples of standard electrical potentials for different elements.

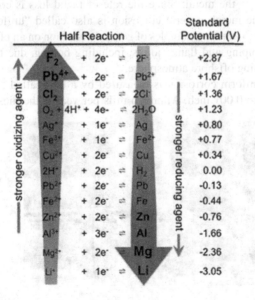

Half Reaction	Standard Potential (V)
$F_2 + 2e^- \rightleftharpoons 2F^-$	+2.87
$Pb^{4+} + 2e^- \rightleftharpoons Pb^{2+}$	+1.67
$Cl_2 + 2e^- \rightleftharpoons 2Cl^-$	+1.36
$O_2 + 4H^+ + 4e^- \rightleftharpoons 2H_2O$	+1.23
$Ag^+ + 1e^- \rightleftharpoons Ag$	+0.80
$Fe^{3+} + 1e^- \rightleftharpoons Fe^{2+}$	+0.77
$Cu^{2+} + 2e^- \rightleftharpoons Cu$	+0.34
$2H^+ + 2e^- \rightleftharpoons H_2$	0.00
$Pb^{2+} + 2e^- \rightleftharpoons Pb$	-0.13
$Fe^{2+} + 2e^- \rightleftharpoons Fe$	-0.44
$Zn^{2+} + 2e^- \rightleftharpoons Zn$	-0.76
$Al^{3+} + 3e^- \rightleftharpoons Al$	-1.66
$Mg^{2+} + 2e^- \rightleftharpoons Mg$	-2.36
$Li^+ + 1e^- \rightleftharpoons Li$	-3.05

(stronger oxidizing agent) (stronger reducing agent)

FIGURE 1.4 Standard electrical potential values for different elements. (Photograph by author.)

Now it is possible to calculate the electrical potential for a cell with zinc as an anode and copper as a cathode. The electrical potential of the cell, according to Equation 1.3, is equal to $0.34 - (-0.76) = 1.10 \, V$. Copper-zinc electrodes are typically used inside batteries to produce electricity.

1.4 OFFSHORE CORROSION TYPES

As discussed above, corrosion is a major problem in the offshore environment due to the extreme corrosivity of the operating and environmental conditions, as well as the presence of corrosive compounds in the fluid service. For the purposes of this chapter, external corrosion mechanisms are divided into three categories and seven subtypes: (1) general corrosion; (2) localized corrosion, such as crevice, pitting and galvanic corrosion, and corrosion under insulation (CUI); (3) environmental or mechanical corrosion cracking, such as chloride stress cracking corrosion (CLSCC) and hydrogen-induced stress cracking (HISC) corrosion.

1.4.1 GENERAL CORROSION

General corrosion, in which the metal corrodes uniformly, is the simplest type of offshore corrosion. General corrosion is common for carbon steel materials, especially if these materials are not coated properly. General corrosion can occur as an external type of corrosion caused by the corrosive offshore environment, and it can also occur internally inside piping and equipment made of carbon steel due to the presence of carbon dioxide (CO_2). General corrosion causes a decrease in the thickness of the metal. Since the rate of metal loss is constant over the entire area of the metal, general corrosion is also called "uniform corrosion." Figures 1.5 and 1.6 illustrate samples of general corrosion on an offshore platform and on valves, piping and flange joints, including bolting, due to the corrosive chloride-containing offshore atmosphere.

General or uniform corrosion is measured by a unit called "mils per year"; 1 mils is equal to 0.001 inch. Although mils per year is the most common unit

FIGURE 1.5 General corrosion of a valve including connected flanges and piping in the offshore environment. (Courtesy: Shutterstock.)

FIGURE 1.6 General corrosion on flange bolts and nuts in the offshore environment. (Courtesy: Shutterstock.)

TABLE 1.2
Categories of General Corrosion Rate

Category	Corrosion Rate	Corrosion Performance
I	<0.15 mm/year (0.006 inch/year) or (6 mils/year)	*Acceptable*: The corroded component at this corrosion rate can perform properly for the given application
II	0.15–1.5 mm/year (0.006–0.06 inch/year) or (6–60 mils/year)	*Medium*: A medium corrosion rate is associated with satisfactory performance of the corroded component. A faster corrosion rate than this category is not tolerable
III	>1.5 mm/year (>0.06 inch/year) or (>60 mils/year)	Not acceptable

of measurement, other units such as millimeter or inch per year are alternative measures of this type of corrosion. The general corrosion rate can be grouped into three categories: low, medium and high, as shown in Table 1.2. Overall, general corrosion is easier to manage and control compared to other types of corrosion.

Figure 1.7 illustrates a valve with a manual operating mechanism consisting of a handwheel plus a gear box. A human operator must rotate the handwheel in order to open or close the valve. The gearbox installed on the top of the valve contains gears inside in order to increase the force applied by the operator and transmit it to the valve internals. As an example, if an operator provides a force of 200 N and the gear box ratio is 10, then ten times of the operator applied force, equal to 2,000 N, is transmitted and applied to the valve internals. The gearbox of the valve in the figure is in cast iron coated with zinc-rich primer and coating system 1 as per the NORSOK M-501 coating standard. NORSOK refers to the standards of the Norwegian offshore industry; coating systems according to the NORSOK standard are explained in more detail in Chapter 3.

The coating of the valve gearbox is peeling off or flaking as a result of adhesion failure. Figure 1.8 shows another gearbox in cast iron coated with zinc epoxy

FIGURE 1.7   Cast iron gearbox corrosion and zinc-rich coating peeling off. (Photograph by author.)

FIGURE 1.8   Epoxy coating flaking off from a cast iron valve gearbox. (Photograph by author.)

primer in an offshore project where the paint is separated from the base metal in cast iron substrate in some areas, resulting in general corrosion of the gear box. "Substrate" in coating terminology refers to the solid, metallic surface that will be coated. Cast iron, like carbon steel, cannot provide high corrosion resistance in the offshore environment and is subject to rust and general corrosion; see the gearboxes illustrated in Figures 1.7 and 1.8. The reason for the coating failure is not completely clear, but the poor adhesion of the coating to the substrate could be due to poor surface preparation. Roughness is an important parameter of adhesion between the coating and the substrate, and cast iron has high surface roughness. A past study (see Reference [11]) highlighted the effect of cast iron microstructure and surface roughness on adhesion and adherence of epoxy protection and con-cluded that decreasing the surface roughness of cast iron during surface prepara-tion improves the adhesion and strength of the epoxy primer.

Although coating the valve gear boxes shown in Figures 1.7 and 1.8 was not completely successful and led to general corrosion in some areas after coating

FIGURE 1.9 General corrosion on a valve handwheel and gearbox made of steel in the offshore environment. (Photograph by author.)

failure or removal, Figure 1.9 illustrates more severe general corrosion on a valve gearbox and handwheel made of uncoated steel in the offshore environment.

What is the solution in this case, where coating the gearbox was unsuccessful? The proposal of the present author, based on industrial experiences, is to change the material of the gearbox and handwheel from cast iron or carbon steel to stainless steel (SS) grade 316. It should be noted that the material of the handwheel in Figure 1.7 is SS316; no sign of corrosion can be seen on the handwheel, unlike the steel handwheel in Figure 1.9. "Steel" is a general name given to a large family of alloys that mainly contain carbon and iron. Cast iron and carbon steel are both considered kinds of steel. The main difference between carbon steel and cast iron is related to the percentage of carbon; carbon steel has <2% carbon, while cast iron has at least 2% carbon. SS316 is a type of austenitic stainless steel made of iron, carbon, chromium, nickel and other alloys. Austenitic stainless steels could be in different grades, such as 304, 316, 321, 347, etc. Among all these grades, SS316 has the highest corrosion resistance in the offshore environment. In general, however, SS316 is not a popular choice of material in the Norwegian offshore industry. The main reason is related to corrosion, particularly the susceptibility of SS316 to chloride in the offshore environment. NORSOK M-001, the material selection standard, does not allow the usage of SS316 in operating temperatures above 60°C due to the risk of pitting corrosion and CLSCC, which are explained in more detail later in this chapter. Valves with SS316 gearboxes could have operating temperature values above 60°C. But why are gearboxes in stainless steel 316 are permitted for such valves with an operating temperature above 60°C? The operating temperature of a valve is associated with the fluid service temperature inside the valve. But the gearbox is not in contact with the fluid service, so the temperature of the gearbox that is in contact with cold offshore atmosphere would be less than 60°C. In addition, a gearbox is not a pressure-containing part like a valve body, and there is no pressurized fluid or stress inside a gearbox. CLSCC requires both chloride contact and stress to occur. Therefore, as shown in

FIGURE 1.10   Gearbox in stainless steel 316 material. (Photograph by author.)

Figure 1.10, a gearbox in SS316 without any coating has been effectively used for manual valves in a couple of recent projects in the Norwegian offshore industry.

1.4.2   Localized Corrosion

Unlike general or uniform corrosion, localized corrosion occurs at specific areas of the surface with more intense attack, and the localized sites corrode faster and more pervasively compared to other areas of the metal. Localized corrosion is categorized into four main types: crevice, pitting and galvanic corrosion and CUI.

1.4.2.1   Crevice Corrosion

Crevice corrosion refers to a localized attack on a metal surface at or near the gaps and crevices between two jointing surfaces. The gap or crevice can be formed between two metals or between a metal and a soft or non-metal material. Crevice corrosion occurs in areas where the fluid becomes trapped, such as under gaskets, washers, insulation materials, fastener heads, threaded connections, etc. Different types of compounds can get trapped in crevices inside piping systems, such as corrosive fluid, dirt, mud, biofouling and other types of deposits. Fouling organisms or biofouling refers to animals or plant species that live in water and can adhere to the surface of materials used in water.

If we think about industrial valves, there are areas where gaps or crevices exist, such as flange faces. Flanging is a method of connecting pipes, valves and equipment to form a piping system, as illustrated in Figure 1.11. As shown in the figure, flanges are connected together with bolts and nuts. In fact, using flange

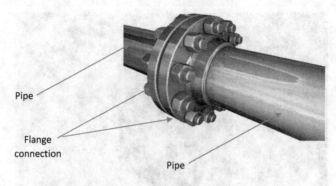

FIGURE 1.11 Flange connection. (Photograph by author.)

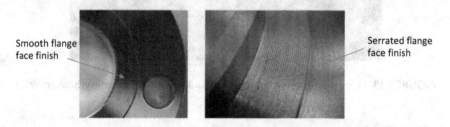

Smooth flange
face finish

Serrated flange
face finish

FIGURE 1.12 Smooth finish vs serrated finish flange face. (Photograph by author.)

joints for connecting pipe, rather than welding two pieces of pipe directly to each other, makes the piping system more expensive but provides the possibility for disassembling the piping system by unscrewing the bolts and nuts to perform repair, cleaning, inspection or modification.

Each flange is connected to a piece of pipe; welding is a common way to connect flanges to piping, but other types of connection, such as threaded, do exist. There is a gasket between two mating flanges, which could be in a metallic, semi-metallic or soft material. A flange face is defined as the surface area that hosts the gasket. A flange face can be smooth or serrated, as illustrated in Figure 1.12. Serrated flange face areas contain some gaps, which are a vulnerable place for crevice corrosion to occur. Austenitic stainless steel is very prone to crevice, pitting and CLSCC. Figure 1.13 illustrates crevice corrosion on a flange face made of austenitic stainless-steel grade 316.

There are five types of flange faces: flat face, raised face, ring-type joint (RTJ), male and female and tongue and groove. Figure 1.14 illustrates a RTJ flange face, which is typically used for high-pressure class piping and valve systems. High-pressure piping is considered for pressure classes of 600 and above, which are equal to a pressure nominal (PN) of almost 100 bar and higher. The groove inside the RTJ flange connection highlighted on the flange on the right corner of Figure 1.14 is a place where crevice corrosion could occur. The metallic RTJ gasket is sitting inside the groove, as highlighted on the left side of Figure 1.14.

FIGURE 1.13   Crevice corrosion on a flange face in 316 material. (Courtesy: Shutterstock.)

RTJ metallic
gasket

Groove inside
the flange face

FIGURE 1.14   Ring-type joint flange face. (Photograph by author.)

Carbon and low-alloy steel materials are not considered corrosion-resistant, so for RTJ flanges in carbon and low-alloy steel material, it is common to apply Inconel 625, which is a highly corrosion-resistant material, on the ring grooves to prevent crevice corrosion. The other two types of corrosion, pitting and CLSCC, can occur in grooves such as those on RTJ flange faces. Thus, applying Inconel 625 on flange ring grooves is a very good way to prevent all three types of localized corrosion. Applying Inconel 625 or other nickel alloys to prevent corrosion on a non-corrosion-resistant alloy is known as "cladding." Inconel 625 is typically clad by weld overlay method in a thickness of 3 mm. Some specifications may require a minimum of two layers of weld overlay on Inconel 625. If the weld overlay is done in sour services, meaning services with a considerable amount of hydrogen sulfide (H₂S), then the hardness of the Inconel 625 should be limited to 35 HRC (Rockwell C) according to ISO 15156 and NACE MR 0175 requirements. These standards address material selection for use in H_2S -containing services in the oil and gas industry. Inconel 625 contains more than 50% nickel, between 20% and 23% chromium and between 8% and 10% molybdenum. This nickel alloy provides high corrosion resistance against various types of corrosion, such as crevice, pitting, hydrogen sulfide and carbon dioxide. The other important properties of Inconel 625 are its high strength and very good weldability and ductility. The strength of Inconel 625 is not of benefit for weld overlay application, because weld overlay is not performed for improving mechanical strength.

The type of weld cladding on the metal surface is gas tungsten arc welding (GTAW) or tungsten inert gas (TIG) welding. ASME IX or ASME Section IX, which is a part of ASME boiler pressure vessel code, contains rules for applying and qualifying the welding procedure and welder and is applicable for Inconel 625 cladding or weld overlay. After applying the weld overlay, the Inconel 625 cladding should be machined to achieve the required thickness, e.g., 3 mm, and a post-weld heat treatment (PWHT) is then applied on the weld overlay to release/reduce the residual stress inside the welding and prevent cracking and brittle fracture in the welded areas. Applying PWHT is a good way to control the hardness of the materials, which is critical in the case of requirements for sour service. The weld overlay could be subject to non-destructive testing (NDT) to make sure that it is without defect. NDT is a testing and analysis technique used in the industry to evaluate the properties of a material, component, structure or system for characteristic differences or welding defects and discontinuities without causing damage to the original part. Different NDT methods, such as visual examination, a magnetic particle test or a liquid penetration test, are used to detect defects on the surface. In addition, volumetric NDTs, such as radiography and ultrasonic tests, are implemented in the industry. Radiography is an examination or inspection NDT technique that is applied on Inconel 625 weld overlays. An X-ray or gamma ray is used in a radiography examination to detect different material defects and discontinuities, such as cracks and voids. A radiography test is based on the differential absorption of the penetrating radiation into the target material.

Alternatively, a laser welding process can be used for cladding; this process provides additional benefits compared to a conventional welding method such as TIG. The first advantage of laser welding is that it is very fast—it can weld meters of steel in a single minute. The second advantage of laser welding is that it is typically done automatically without any need for an operator. Figure 1.15 illustrates a robotic or automatic laser welding machine used in the valve industry. Laser welding provides very narrow and uniform, deep, high-quality welding.

The other advantage of laser welding is that it produces less thermal input, which causes less thermal damage to the welded metal. The low heat input from laser welding prevents weld distortion. Weld distortion refers to any unwanted physical change of the base metal. The other positive point with regard to laser welding is related to the low substrate dilution of the welding. Weld dilution refers to a change in the composition of the weld metal caused by mixing with the base

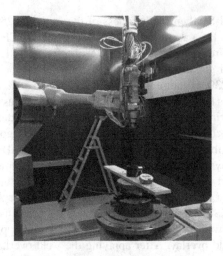

FIGURE 1.15   Laser welding machine in the valve industry. (Photograph by author.)

FIGURE 1.16   Uniform, deep, high-quality laser welding around the disk of a butterfly valve. (Photograph by author.)

metal. Dilution of the base metal to the weld overlay could be limited to 5% or 10%, depending on the fluid service and welding specification requirements; using laser welding assures that the limitation requirement is satisfied.

Figure 1.16 illustrates high-quality, uniform laser welding of the outer part of the disk of a butterfly valve. Butterfly valves, which are explained in Chapter 5, are a type of on/off valve used to start and stop the fluid in a piping system. Butterfly valves are widely used in the Norwegian offshore industry for on/off applications in utility services. Utility services involve non-corrosive, low-pressure class fluids such as water, air, etc.

Inconel 625 weld overlay can be done on serrated flange faces in carbon steel to prevent crevice corrosion, as illustrated in Figure 1.17. Weld cladding is not limited

FIGURE 1.17 Inconel 625 cladding or weld overlay on a carbon steel serrated flange face. (Photograph by author.)

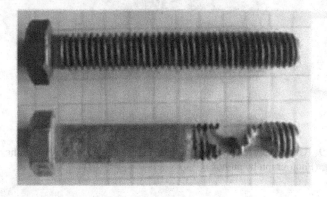

FIGURE 1.18 Crevice corrosion in the threaded areas of a bolt in the offshore environment after 5 years. (Photograph by author.)

to piping flanges. Piping and valve internals that are made in carbon or low-alloy steel can be clad with Inconel 625 to increase the piping material corrosion resistance and material design life. Using a weld overlay of Inconel 625 on carbon steel is a cheaper and more cost-effective solution compared to using solid Inconel 625. In addition, valve sealing grooves in carbon and low-alloy materials are typically treated with a weld overlay of Inconel 625 to prevent crevice corrosion.

Bolts and nuts in piping systems are used to secure flange connections on piping and valves. Bolts and nuts, also called fasteners, have threaded areas that are at high risk of crevice corrosion. Crevice corrosion occurs at the threads and cavities of fasteners that are not ventilated properly, and in areas exposed to corrosive fluid and/or moisture, which become stagnant and act as an electrolyte. Fasteners are typically low-cost components, but their failure can be very catastrophic and costly, and render the connected flanges and valves useless. Figure 1.18 illustrates

FIGURE 1.19 Crevice corrosion in the threaded areas and head of a bolt in the offshore environment. (Photograph by author.)

crevice corrosion on the threaded connections of a bolt that was used in the offshore environment for 5 years. Figure 1.19 illustrates rust and crevice corrosion on a bolt head and threads.

1.4.2.2 Pitting Corrosion

Pitting is a form of extremely localized attack that results in holes and pits inside the metal. Pitting is the classic type of corrosion that occurs when metals are exposed to seawater or humid environments containing salt and chloride. The holes that form inside the metal due to pitting corrosion may be either large or small and are not uniform. In fact, pitting corrosion is known as one of the most destructive forms of corrosion, as it is hard to predict, detect and characterize. In this type of corrosion, either an anodic point or more commonly a cathodic point forms a small corrosion cell with the surrounding environment. It first appears as gray and white powdery deposits similar to dust, which form blotches on the metal surface. When these deposits are cleaned away, tiny pits or holes can be seen on the surface of the metal. Pitting corrosion starts with the formation of pits, which can grow into holes and cavities. The pits typically penetrate from the surface and move downward vertically. Figure 1.20 illustrates pitting corrosion on a metal surface. The geometry and shape of the pits are very similar to those created in crevice corrosion. The difference is related to the location of the pits, which are not formed in crevices and grooves in pitting corrosion, unlike crevice corrosion.

Damage to the protective film on a metal surface and/or non-uniformities in a metal structure can increase pitting corrosion speed. One of the main conditions for pitting to occur is material vulnerability. Aluminum alloys and austenitic stainless steels, such as grades 304, 321 and 347, are at high risk of pitting corrosion. These susceptible materials have an oxide film on their surface, known as a passive film, that protects them from corrosion attack; damage to the oxide film layer can result in pitting corrosion initiation. Figure 1.21 illustrates pitting corrosion in austenitic stainless steel 316 caused by the removal of the protective oxide layer in the offshore environment.

The second important factor that contributes to pitting corrosion is the presence of chloride in the offshore environment. Temperature and the presence and

FIGURE 1.20 Pitting corrosion on a metal surface. (Courtesy: Shutterstock.)

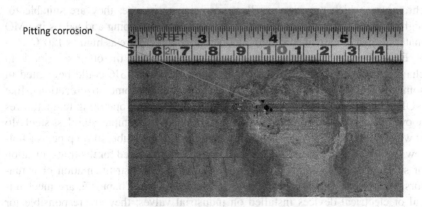

FIGURE 1.21 Pitting corrosion in stainless steel 316 after removal of the protective layer in the offshore environment. (Courtesy: Shutterstock.)

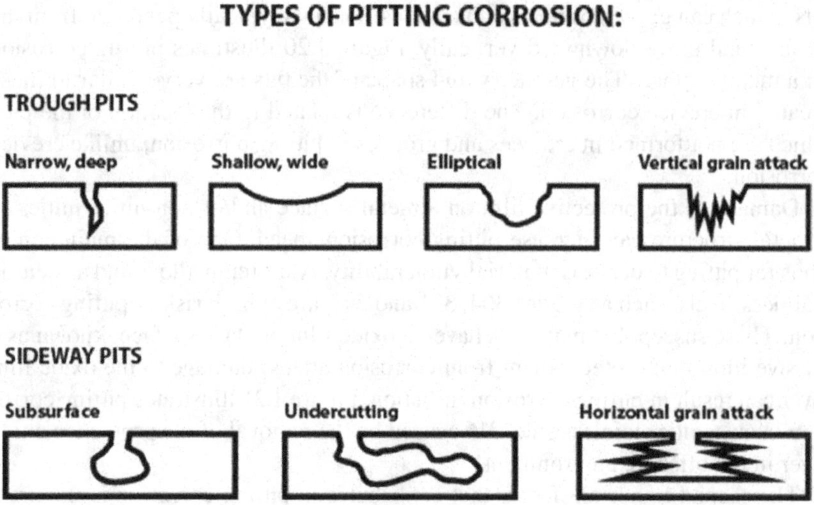

TYPES OF PITTING CORROSION:

TROUGH PITS

Narrow, deep   Shallow, wide   Elliptical   Vertical grain attack

SIDEWAY PITS

Subsurface   Undercutting   Horizontal grain attack

FIGURE 1.22   Types of pits in metal due to pitting corrosion. (Photograph by author.)

concentration of oxygen increase the severity of pitting corrosion. Different types of pitting corrosion are shown in Figure 1.22.

Coating can be used in some cases to prevent external pitting corrosion in the offshore environment. For example, it is common to coat piping and valves in duplex material if they are used for operating temperatures above 100°C. The maximum temperature in which a super duplex material can be used offshore without coating is 110°C. Duplex and super duplex are families of stainless steels that have dual properties: austenitic and ferritic. Stainless steels are iron-based alloys with at least 11% chromium content. Austenitic stainless steels are those with grades in the 300 series, such as 316; austenitic stainless-steel materials contain iron, chromium, nickel and minor alloys. Ferritic stainless steels contain 10%–30% chromium, which provides excellent corrosion resistance; they are suitable for high-temperature applications and have good strength. Piping and valves in 6MO materials can remain uncoated for an operating temperature as high as 120°C.

Besides applying coating to prevent pitting corrosion, the other solution is to change the material. Although austenitic stainless steel 316 could be coated in some projects to prevent pitting and chloride attack, some specifications, like NORSOK, do not allow the use of stainless steel 316 in operating temperatures above 60°C, even with coating implementation. As an example, stainless steel 316 is widely used for tubing in the offshore environment. A tube, like a pipe, is a hollow cylinder used to move fluid. Tubes are typically selected for the transportation of substances, such as air and hydraulic oil, used in instrumentation or actuators. Actuators, which are explained in more detail in Chapter 5, are mechanical or electrical devices installed on industrial valves; they are responsible for automatic, operator-less movement of the valves between opening and closing positions. Mechanical actuators on topside platforms work with air or hydraulic

Air filter and
pressure gauge

Ball valve

FIGURE 1.23 Tubing system to supply air to pneumatic actuators. (Courtesy: Shutterstock.)

oil. Routing tubing is easier than routing pipe, since tubing is more flexible. In addition, lengths of tubing, unlike piping in an offshore topside plant, are not welded together, so there is no need to apply welding and NDT on tubing joints. Figure 1.23 illustrates a tubing system used in an offshore plant to transport air to pneumatic actuators. The figure shows many small tubes, an air filter and regulators, as well as small ball valves, which are all considered part of the control system. The air filter removes sand from the air since the cleanliness of the air is important for proper actuator functioning. Particle contamination in the air and/or hydraulic fluid is the main cause of actuator failure. The air filter is integrated with a pressure gauge to measure the air pressure. A ball valve is installed on the tubing system to start and stop the airflow.

Stainless steel 316 tubing is at risk of crevice and pitting corrosion, as illustrated in Figure 1.24, so in some cases, the tubing may be upgraded to a more corrosion-resistant material like 6MO.

6MO (UNS S31254) is a super austenitic stainless steel with a relatively high level of molybdenum (6%) and nitrogen, which provides a high level of pitting and crevice corrosion resistance. The resistance of a material to pitting and crevice corrosion is measured by pitting resistance equivalent number (PREN). There are several formulas for PREN calculation, but the simplest one correlates the PREN to three elements in the material: chromium, molybdenum and nitrogen. Equation 1.4 expresses this basic PREN calculation.

PREN calculation

$$PREN = Chromium(Cr) + 3.3 * Molybdenum(Mo) + 16 Nitrogen(N) \quad (1.4)$$

FIGURE 1.24 Crevice and pitting corrosion of tubes in the offshore environment. (Photograph by author.)

Stainless steel 316 contains 17% chromium, 2.5% molybdenum on average and at maximum 0.10% nitrogen. Using the average values for chromium and molybdenum and the maximum value of nitrogen in Equation 1.4 gives the PREN for stainless steel 316 as follows:

$$Stainless steel 316 PREN = 17 + 3.3 * 2.5 + 16 * 0.10 = 26.85$$

22Cr duplex stainless steel has higher crevice and pitting corrosion resistance compared to stainless steel 316. 22Cr duplex contains 22% Cr, 3% Mo and 0.18% N. Using the provided chemical composition values, the PREN for 22Cr duplex is calculated as follows:

$$Stainless steel 22 Cr duplex PREN = 22 + 3.3 * 3 + 16 * 0.18 = 34.78$$

25Cr super duplex stainless steel has higher crevice and pitting corrosion resistance compared to stainless steel 316 and 22Cr duplex. 25Cr duplex contains 25% Cr, 3.4% Mo and 0.17% N on average. Using the provided chemical composition values, the PREN for 25Cr duplex is calculated as follows:

$$Stainless steel 25 Cr super duplex PREN = 25 + 3.3 * 3.4 + 16 * 0.17 = 38.94$$

6MO has higher crevice and pitting corrosion resistance compared to stainless steel 316, 22Cr duplex and 25Cr super duplex. 6MO contains 20.6% Cr, 6.3% Mo and 0.21% N on average. Using the provided chemical composition values, the PREN for 6MO is calculated as follows:

$$6MO PREN = 20.6 + 3.3 * 6.3 + 16 * 0.21 = 44.75$$

The other material vulnerable to pitting corrosion in the offshore environment is 17-4PH stainless steel. "PH" in front of the material name stands for precipitation hardening heat treatment process. The high strength of 17-4PH comes from its heat treatment. Precipitation hardening, also called age hardening, is a process in which the hardness and mechanical strength of a material are enhanced significantly by the formation of extremely small, uniformly distributed particles of a second phase within the original phase matrix, in this case, copper participation. 17-4PH is a very hard martensitic stainless steel containing about 17% chromium, 4% nickel and 4% copper. It is selected in onshore plants, like petrochemical plants and refineries, for valve stems or shafts due to its high strength. However, 17-4PH is at high risk of pitting and SCC in chloride-containing environments. Figure 1.25 illustrates pitting and SCC on a 17-4PH shaft in the offshore sector caused by contact with chloride. Thus, it is not recommended to select 17-4PH for valve shafts in the offshore environment. The alternative shaft or stem material in offshore is Inconel 718. Inconel 718, like 17-4PH, is a very strong and hard material that has undergone age-hardening heat treatment. CLSCC is explained in the next subchapter of this section.

Martensitic stainless steels, also known as series 400 stainless steels, contain 11%–17% chromium and have high mechanical strength and hardness due to age-hardening heat treatment. They are mainly used for turbines and razor blades due to their mechanical and wear resistance. The corrosion resistance of martensitic stainless steels is less than other types of stainless steels. Martensitic stainless-steel grade 410, known as 13% chromium, is a common stem material for carbon steel body valves in onshore refineries and chemical plants. Martensitic stainless-steel grade 416, known as 13% chromium–4% nickel, is a common stem material for carbon steel body valves in the topside offshore sector on platforms

FIGURE 1.25 Pitting and chloride stress cracking corrosion of a 17-4PH shaft. (Courtesy: Shutterstock.)

and ships. However, 13Cr-4Ni or stainless steel 416 is not recommended for sub-sea applications based on the present author's experience.

1.4.2.3 Galvanic Corrosion

Galvanic corrosion, also called *dissimilar metal corrosion* or *bimetallic corrosion*, is another type of localized corrosion. It occurs when two dissimilar materials are coupled in a corrosive electrolyte, so it is a type of electrochemical corrosion. There are three important parameters in galvanic corrosion: different types of metals, the presence of an electrolyte and electrical continuity between the two metals. Since this chapter is dedicated to offshore corrosion, it should be noted that the humid, offshore environment plays the role of an electrolyte. Galvanic corrosion is initiated by the presence of an electrical cell, which was explained above. Figure 1.26 illustrates the coupling of a valve stem key in low-alloy steel 4140 and a valve stem in 22Cr duplex in a humid, offshore environment. As shown in the figure, the less noble metal, which is the low-alloy steel stem key in this case, acts as an anode, loses electrons and becomes corroded. The 22Cr duplex stem material is more noble; it acts as a cathode and is protected against corrosion. The stem of a valve is a pressure-containing part that connects the valve operator to the valve closure member. The valve closure member is an internal component that moves inside the valve to close or open it, or to regulate the flow inside the valve. The stem and stem key shown in the picture belong to a ball valve. A ball valve, as explained in more detail in Chapter 5, is only used for opening and closing applications or to start/stop the fluid. Pressure-containing parts of valves are components, such as the valve stem, whose failure to function leads to leakage from the valve to the environment. The stem key is attached to the stem to show the opening or closing position of the valve when the gearbox or lever is removed from the top of the valve. Typically, the opening and closing positions of the valves are shown on the gearbox and lever. However, the gearbox and lever could be disassembled from the valve during maintenance. In such cases, the stem key stands parallel to the hole inside the closure member in the ball valve, which enables the operator or other personnel working with the valve

Galvanic corrosion on the stem key in low alloy steel

Stem of the valve in 22Cr duplex

FIGURE 1.26 Galvanic corrosion on the stem key of a valve. (Photograph by author.)

to ascertain the position of the valve. During normal operating conditions, the stem key and the upper part of the stem attached to the stem key are covered by the valve gearbox, so they are not exposed to the corrosive offshore environment. However, when the gearbox is removed from the stem top, the upper part of the stem and the connected stem key are exposed to the environment and galvanic corrosion takes place.

The other example of galvanic corrosion in the offshore industry associated with butterfly valves has to do with contact between graphite and an active metal in a corrosive environment. Butterfly valves are a type of valve used to stop/start fluid such as seawater in the Norwegian offshore industry. Seawater contains chloride and is known as a corrosive service that acts as an electrolyte. The valve stem area is filled with soft materials called stem sealing or packing. Graphite packing is common for the stem sealing of industrial valves, as graphite can be used for a wide range of temperatures, in high-pressure applications and in corrosive environments. Figure 1.27 illustrates a valve stem with three graphite packing layers around the stem sealing. If the valve stem is selected in a less noble grade of stainless steel, such as 13Chromium or 22Chromium stainless steel grades, these stem materials are prone to galvanic corrosion in contact with the graphite packing and corrosive seawater media. This type of galvanic corrosion is internal, as it is caused by the fluid service, unlike the previous example. The solution that has been proposed, accepted and implemented by material and valve engineers to prevent galvanic corrosion between graphite packing and stainless-steel stems in the presence of seawater is to isolate the graphite packing with a lip seal; one lip seal is placed at the bottom of the graphite packing to prevent the seawater service fluid from coming into contact with the packing and thus prevent galvanic corrosion, as illustrated in Figure 1.28. The lip seal shown in white in Figure 1.28 is made of a soft, thermoplastic material, such as Teflon, which is energized with a metallic spring in Inconel 625 material in this case.

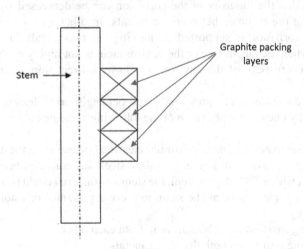

Stem

Graphite packing layers

FIGURE 1.27   Valve stem with three packing layers. (Photograph by author.)

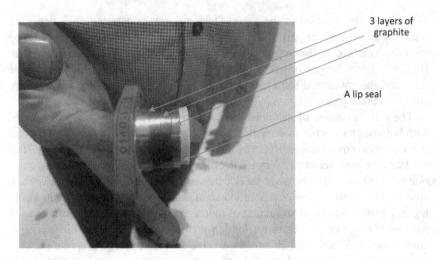

FIGURE 1.28   Valve stem sealings, including three layers of graphite and one lip seal. (Photograph by author.)

Different parameters affect the severity and rate of galvanic corrosion. One is the difference between the electrical potential of the anode and cathode; a higher electrical potential difference between the anode and cathode increases the galvanic corrosion rate on the anode. Another important factor in galvanic corrosion is related to the ratio of the anode and cathode surfaces. If the anode area is larger than the cathode area, the corrosion is negligible. On the other hand, a small ratio of anode to cathode area is undesirable as it results in very rapid corrosion of the anode. The other important parameter in galvanic corrosion is that it is limited to the contact zones between two dissimilar metals. This means that the intensity of the corrosion can be decreased significantly by increasing the distance between the metals. In other words, the decrease in galvanic corrosion is supported by moving the two metals further apart, especially when the electrolyte or the environment is not highly corrosive. The nature and corrosivity of the environment have a direct impact on galvanic corrosion.

Galvanic corrosion can be prevented by proper engineering design, which can be achieved by one or a combination of the following strategies:

1. Coupling materials that have similar electrical potential; in the example where the valve stem key in low-alloy steel was corroded because of contact with a 22Cr duplex stem, the stem key material could be selected in 22Cr duplex, same as the stem, to prevent galvanic corrosion of the stem key.
2. Keeping two dissimilar metals away from each other;
3. Applying coating on both dissimilar metals;
4. Adding corrosion inhibitor to the metal susceptible to being corroded.

1.4.2.4 Corrosion under Insulation (CUI)

Piping and valves in the offshore oil and gas industry can be insulated with non-metallic blanket materials, such as silica aerogel (trade name Pyrogel) reinforced with glass fiber. In addition, metallic isolation boxes can be used around valves and flanges for some isolation classes to provide fire or acoustic protection. There are various reasons for piping and valve isolation, such as heat conservation, cold conservation, personnel protection in cases of operating temperatures higher than 70°C or lower than −10°C, frost protection in low-operating temperatures (e.g., 10°C or less), ice and condensate protection, fire protection and acoustic insulation. Some insulation classes or materials are used for multiple purposes, such as heat conservation and acoustic isolation, heat conservation and fire protection or various other combinations. Figure 1.29 illustrates a partially insulated piping assembly. Figure 1.30 illustrates insulation with the purpose of heat conservation and personnel protection around a valve used in a hot process fluid service on an offshore platform. Alternatively, valves and flanges can be insulated with metallic boxes, as illustrated in Figure 1.31. The boxes should be removable to provide access to the isolated valve for inspection and maintenance. Figure 1.32 illustrates the formation of ice on piping and connected valves due to very cold temperature (cryogenic temperature) inside the piping and valve, coupled with high-pressure drop inside the valve. The formation of ice damages both the valve and the icing protection insulation.

CUI is a severe type of localized corrosion that occurs under the insulation of insulated carbon, low-alloy and stainless steels, especially on piping, valves and pressure vessels. It is mainly caused by the penetration of water. Although CUI is categorized as localized corrosion in this book, it can also be considered general corrosion. Figure 1.33 illustrates a Y-type strainer with CUI. A strainer is a piping component that is installed before pumps and compressors to filter out and prevent the ingress of particles into pumps and compressors. CUI causes many cases

FIGURE 1.29 Piping assembly with partial insulation. (Courtesy: Shutterstock.)

FIGURE 1.30   A valve insulated with the double purpose of heat conservation and personnel protection in a hot process service. (Courtesy: Shutterstock.)

FIGURE 1.31   Valves insulated with boxes in the offshore sector. (Courtesy: Shutterstock.)

of leakage in piping and valves and a considerable amount of maintenance cost. The main risk associated with this type of corrosion is related to the difficulty of detecting it, as it is hidden under the insulation and thus cannot be seen in normal operating conditions. CUI is a major corrosion problem worldwide and can occur in both offshore and onshore sectors of the oil and gas industry.

CUI is typically caused by water seeping under the insulation through a break in the insulation. The water or moisture remains under the insulation and corrodes the metal surface until the insulation is removed for inspection. The main

FIGURE 1.32   Ice formation on piping and a valve, damage to the insulation and the valve and CUI. (Courtesy: Shutterstock.)

source of the water could be rain, process liquid water leakage or deluge systems. A deluge system is the main firefighting system in the offshore oil and gas industry; it supplies water to extinguish fires. The CUI initiated by water can be intensified by the presence of contaminants such as chloride and sulfide. As noted above, the offshore environment contains chloride, which can increase the corrosivity of the water trapped under the insulation. If the corrosivity of the water is intensified by both chloride and stress, then the CUI can be converted to CLSCC for vulnerable materials such as austenitic stainless steels. CLSCC is explained later in this chapter. The source of the chloride or sulfide contaminations could be the insulation itself. Insulation can increase the severity of corrosion in other ways too, such as providing an annular space or crevice for water to accumulate due to its permeability and possible retention and absorption of the water, or by providing other contaminates that could increase the corrosion rate, such as those that can react with water and make acid.

The corrosion rate of CUI could be between 1.5 and 3 mm per year, which could be from 20 times up to almost 40 times the rate of corrosion in the offshore environment. The other interesting statistics is that on average, 60% of the insulation types in the oil and gas industry that have been in operation for more than 10 years contain CUI due to water and/or moisture accumulation. Temperature is another parameter that affects the criticality of CUI. Higher temperature in general increases the corrosion rate of CUI—but not always. The risk of CUI for carbon steel material is normally in the temperature range between −4°C and +175°C, but the highest risk of corrosion is in the temperature range between +60°C and 120°C. For stainless steel material, the risk of CUI is in the temperature range between +50°C and 175°C. The other important parameter that increases the CUI corrosion rate is the availability of oxygen.

Which areas of the piping system are more prone to CUI? Low elevated points or low points where water can accumulate. As an example, the bottom of the

Y-type strainer shown in Figure 1.33 is a low point where water can be trapped. The other areas that are vulnerable to CUI are welded joints and places where the pipe changes direction through a bend or elbow. There are two reasons why pipe welding and bends are vulnerable to CUI; first, it is difficult to wrap insulation on welded and rotated areas, and second, these areas are mechanically weak. Figure 1.34 illustrates wrapped insulation damage and removal, as well as rust and corrosion, in an area where a pipe is rotated 90°. Figure 1.35 shows piping with thermal isolation. The insulation is cracked and damaged in the area close to the weld seam, which provides a suitable place for water to become trapped and cause CUI. The cause of the crack in the insulation piping is unclear, but it could be due to excessive vibration and fatigue stress in the piping system. Fatigue refers to the cracking of materials because of cyclic or fluctuating levels of stress. The

FIGURE 1.33   Corrosion under insulation on a Y-type strainer. (Courtesy: Shutterstock.)

FIGURE 1.34   Damage to wrapped insulation and CUI at a 90° bend in a pipe. (Courtesy: Shutterstock.)

FIGURE 1.35 Damaged thermal insulation close to a piping weld seam. (Courtesy: Shutterstock.)

Termination of insulation close to flange connection

FIGURE 1.36 Termination of insulation and porous insulation close to a flange connection. (Courtesy: Shutterstock.)

loads that cause fatigue could be tensile, compression, torsion or thermal. The other area where there is a high risk of insulation damage and CUI is the insulation termination points. Figure 1.36 illustrates the termination of insulation close to a flange connection where the insulation is porous and ready to absorb water.

The other areas where insulation is at high risk of damage and CUI are pipe branches and pipe-supported areas. Figure 1.37 illustrates the damage of piping cold conservation insulation on a piping branch connection.

There are two main approaches to preventing CUI; the first is to avoid water ingress into the insulation materials. Three main tasks should be performed to avoid ingress of water into the insulation and thus onto the steel surface; the first

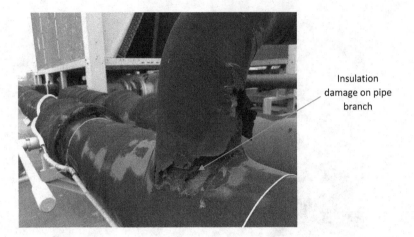

Insulation damage on pipe branch

FIGURE 1.37 Damaged cold conservation insulation and CUI on a pipe branch area. (Courtesy: Shutterstock.)

is to choose suitable insulation materials. The second is to apply the insulation correctly around the insulated item, and the third task is to protect the insulated items against rainwater. The second approach to preventing CUI is to apply coating to the material prior to installing the insulation. Chapters 3 and 6 will discuss types of coating under insulation in more detail.

1.4.3 ENVIRONMENTAL CRACKING CORROSION

1.4.3.1 Chloride Stress Cracking Corrosion

CLSCC is a common type of corrosion for austenitic stainless steels. CLSCC is a specific type of SCC caused by chloride. SCC can be caused by other types of media, such as amine and caustic media. For SCC to occur, three conditions must be met: corrosive environment, vulnerable material and tensile stress (see Figure 1.38).

Tensile loads could be either applied or residual or a combination of both. Residual stress refers to loads that remain inside the material even in the absence of external loads. As an example, welding that causes thermal variation in the material can produce residual stresses. Thus, SCC occurs in welding areas in many cases due to residual stress. Figure 1.39 illustrates SCC and leakage in the area where austenitic stainless steels have been welded together.

The initiation of a crack in this type of corrosion could start from corrosion pits or cyclic loads (also called a fatigue crack in that case) or from welding or machining defects. The transformation of pits to cracks depends on many parameters, such as pit depth and shape, and the amount of stress to which the material is subjected. Corrosion caused by a combination of fatigue load and a corrosive environment is called fatigue corrosion. As an example, Figure 1.40 illustrates the shaft of a pump in 17-4PH material that was subjected to constant load cycles

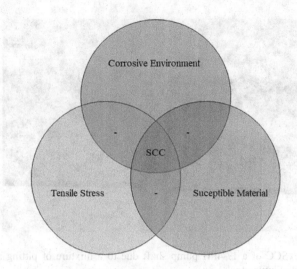

FIGURE 1.38   Contribution of parameters for SCC to occur. (Photograph by author.)

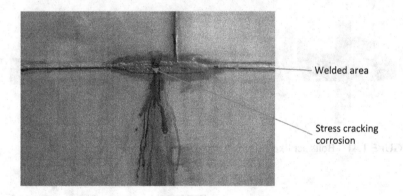

FIGURE 1.39   SCC on the welded area of an austenitic stainless steel. (Courtesy: Shutterstock.)

and fatigue stress. The shaft was in contact with seawater, so the combination of pitting and fatigue resulted in the fatigue corrosion and CLSCC.

Bolts and nuts, also called fasteners, are pipe bulk items used on flange connections. They are explained above in the "crevice corrosion" section. Bolts or fasteners act like a spring, as illustrated in Figure 1.41, meaning that tightening the fasteners provides tension load, and untightening the bolts leads to bolt compression. Bolts can be tightened on flange connections by means of a torque or tension tool. Tightening the bolts with a torque tool can be done either manually or with a hydraulic torque tool. A torque tool is typically used for bolt tensioning in sizes of 1″ or less. Bolt tensioning is another approach to fastening bolts that is common for bolts larger than 1″. Figure 1.42 illustrates a hydraulic torque tool on the left and a manual torque tool on the right for fastening flange stud bolts.

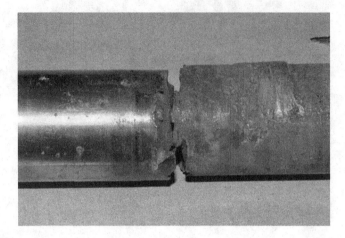

FIGURE 1.40 SCC of a 17-4PH pump shaft due to a mixture of pitting and fatigue stress. (Courtesy: Shutterstock.)

FIGURE 1.41 Bolts act like a spring. (Photograph by author.)

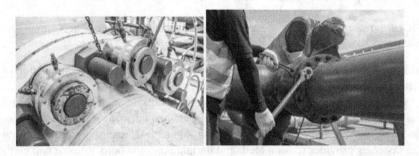

FIGURE 1.42 Tightening flange stud bolts and nuts with hydraulic torque tools (a) and a manual torque tool (b). (Courtesy: Shutterstock.)

FIGURE 1.43   Stud bolt CLSCC in the offshore environment. (Courtesy: Shutterstock.)

Fastening the bolts with a torque or tension tool applies tensile force to the fasteners, which is a contributing factor in fasteners cracking due to SCC.

Austenitic stainless steel bolting under tension in the offshore environment is prone to CLSCC. Figure 1.43 illustrates chloride stress cracking of a stud bolt used for flange connection in the offshore environment. Increases in temperature, concentration of chloride or salt and humidity exacerbate the effects of CLSCC.

1.4.3.2   HISC Corrosion

HISC is a type of corrosion that occurs only in the subsea environment. Like SCC, three parameters are required for HISC to happen; tensile stress and vulnerable materials are two of the HISC occurrence conditions, which are similar to the parameters necessary for SCC occurrence. The third parameter for the initiation of HISC corrosion is a source of hydrogen.

Let's start from the first condition: tensile stress. The type of tensile stress could be applied or residual or a combination of both. As explained above, residual stresses are those that are locked inside a material due to mechanical or thermal loads. As an example, welding can cause thermal stress as well as residual stress inside a material. A heat-affected zone (HAZ) is defined as an area adjacent to welding that is not melted, but in which the microstructure and properties of the material are altered because of the welding. HAZ areas are shown in Figure 1.44 around the welding and are different in color from the base metal. Many crack failures happen in HAZs close to welding areas when the stress value exceeds certain limits. The outcome of stress is strain, which is defined as the ratio of a material's expansion to its initial length. Strain is another parameter that is taken into account in the HISC study and evaluation.

Det Norske Veritas (DNV), a Norwegian company, has developed a standard or guideline addressing HISC for duplex and super duplex materials. The distance

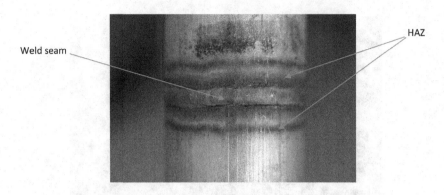

Weld seam

HAZ

FIGURE 1.44   Heat-affected zone areas around welding. (Courtesy: Shutterstock.)

from the weld where HISC poses a higher risk of attack, parameter L_{Res}, is calculated through Equation 1.5.

Calculation of distance from weld joint where HISC could have a higher risk of attack

$$L_{\text{Res}} = 2.5\sqrt{Rt} \qquad (1.5)$$

where:
  R=Nominal pipe radius (mm or inch);
  t=Pipe wall thickness (mm or inch)

Other factors that increase stress in a material and thus increase the risk of HISC attack include sharp corners, grooves and changes in section, size or direction. In addition, using fillet welds, i.e., connecting two pieces of metal together when they are perpendicular in angle, is not recommended in the DNV guideline. Figure 1.45 illustrates cracking occurrence in vulnerable areas of piping.

According to the DNV guideline, materials with higher mechanical strength or, more precisely, higher specified minimum yield strength (SMYS) are at a lower risk of HISC corrosion. But it is important to consider the effect of operating temperature on SMYS. In general, SMYS is decreased by increasing the temperature. Decrease in SMYS due to high temperature can increase the risk of HISC attack according to the stress model provided in the DNV standard; this will be discussed in more detail later in this chapter. The DNV guideline clearly states that the risk of HISC can be reduced by increasing the temperature, however, which is in contradiction with the effect of temperature on SMYS. Thus, separate research is required to clarify the effect of temperature on HISC occurrence.

The second important factor for HISC to occur is susceptible material. Duplex and super duplex stainless steel materials are well known for their high chance of failure because of HISC attack. As a result, DNV has developed a standard and guideline for the design of duplex components, based on DNV-RP-F112, to prevent HISC. However, the risk of HISC attack is not limited to duplex and super duplex.

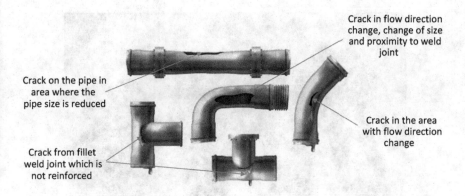

Crack in flow direction change, change of size and proximity to weld joint

Crack on the pipe in area where the pipe size is reduced

Crack in the area with flow direction change

Crack from fillet weld joint which is not reinforced

FIGURE 1.45 Cracks in piping in areas of increased stress. (Courtesy: Shutterstock.)

Hard materials, such martensitic stainless steels, including 13%Cr and 17-4PH, are vulnerable to HISC corrosion attack. In addition, low-alloy and carbon steel bolting, according to API 20E standards such as American Society for Testing and Materials (ASTM) A193 B7, A320 L7, L43 and L7M bolts and ASTM A194 Gr. 4, 7 or 7M nuts, have been failed due to HISC corrosion in many subsea projects (see Figure 1.46). Hard nickel alloys such as Inconel 718 and 725 are among the materials that are at relatively high risk of HISC attack. Since the DNV standard just covers duplex and super duplex, there is a lack of guidance with which to evaluate HISC attack risk and failure analysis for the types of materials mentioned above.

The third and last condition for HISC occurrence is a source of hydrogen. The main source of hydrogen in HISC corrosion is related to cathodic protection. Structures, valves and piping components in subsea environment are at a high risk of external corrosion. Therefore, cathodic protection can be applied to prevent external corrosion. But cathodic protection can lead to the production of hydrogen and cause HISC in vulnerable materials. As discussed above, corrosion happens when a metal loses electrons. Thus, the corrosion of piping, valves or structures in the subsea environment can be prevented by sending the electrons to them. The process of sending electrode from an anode to a protected component or structure is called cathodic protection, as illustrated in Figure 1.47. The sacrificial anode is typically made in a very active metal, such as zinc, magnesium or aluminum, in order to be corroded easily and produce electrons to protect the pipeline, valve or structure, which acts as a cathode.

Assuming zinc in the anode, the zinc reaction to produce electrons is given in Equation 1.6 as follows:

Zinc corrosion in the anode

$$Zn \rightarrow Zn^{2+} + 2e^- \tag{1.6}$$

Extra electrons released from the anode are transferred to the cathode, and the electrons react with oxygen and water in the cathode according to Equation 1.7.

FIGURE 1.46 HISC in low-alloy steel fasteners used in the subsea sector. (Photograph by author.)

FIGURE 1.47 Subsea pipeline cathodic protection. (Courtesy: Shutterstock.)

Reaction in the cathode

$$H_2O + \tfrac{1}{2}O_2 + 2e^- - 2OH \tag{1.7}$$

The production of OH⁻ increases the pH of the seawater and results in the reaction of the OH⁻ with magnesium or calcium to make a protective layer of calcium carbide ($CaCo_3$). The deposition of calcium carbide protects and increases the life of the cathode. However, if the electrical potential of the anode is highly negative, then extra electrons are produced in the anode and are transmitted to the cathode. The extra electrons generated in the anode can cause a second reaction in the cathode, shown in Equation 1.8, that creates a hydrogen source for the HISC attack.

Second reaction in the cathode is due to the generation of extra electrons in the anode

$$H_2O + e^- \rightarrow H + OH^- \qquad (1.8)$$

The other important factor that affects HISC is the material microstructure. At the time of this writing, the mechanism of HISC is not fully and clearly understood. However, in general, it is believed that because hydrogen atoms are small compared to most metallic atoms, they can diffuse between the metallic matrix grain boundaries. What is a metal grain? When a metal solidifies from the molten state, millions of tiny crystals grow and form the grains. Figure 1.48 illustrates the attack of hydrogen atoms on the metal grain boundaries.

The diffusion of the hydrogen into the grain boundaries causes material failure according to two mechanisms and theories; the first is called hydrogen embrittlement local plasticity (HELP) and the second is hydrogen enhancement

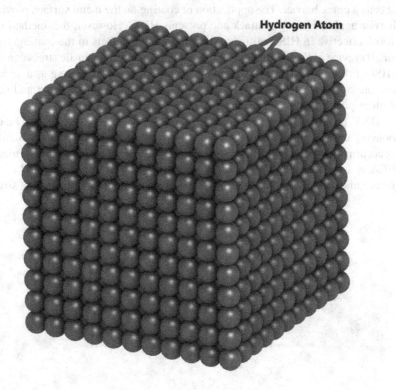

Hydrogen Atom

FIGURE 1.48 Attack of a hydrogen atom on metal grain boundaries. (Courtesy: Shutterstock.)

de-cohesion (HEDC). According to HELP theory, hydrogen atoms are diffused into the void spaces between metal grains and bond together there. The atomic hydrogen causes dislocation in specific metal grains and plastic deformation, which eventually causes a crack in the material. Plastic deformation refers to permanent change in the shape of a material. In HEDC theory, hydrogen enlarges the atoms and molecules, which can then be fractured more easily by the application of stress. According to hydrogen cracking or embrittlement theory, materials with fine grins are at a lower risk of hydrogen attack compared to metals with coarse grains. One proven fact is that the susceptibility of a material to HISC corrosion is higher when the grains are perpendicular to the stress.

There are two types of cracks that can form in a material: ductile and brittle. Ductile cracks form more slowly and result in plastic deformation. A brittle fracture is a very rapid-forming crack that occurs without plastic deformation. A HISC crack is brittle and fast and occurs without any material deformation. Figure 1.49 illustrates hydrogen cracking or embrittlement in the form of a brittle crack in a piece of pipe. The reason why duplex and super duplex are susceptible to HISC is related to the ferritic phase inside these two materials. Duplex and super duplex have two phases: austenitic, which is ductile; and ferritic, which is brittle. Hydrogen makes the ferritic phase embrittle, while the austenitic phase acts as a crack barrier. The application of coating on the metal surface provides a barrier against hydrogen attack and prevents HISC. However, this method is not 100% effective in HISC mitigation due to possible defects in the coating. Even a small crevice in the coating where the metal is under high tensile stress can cause HISC. For this reason, the DNV standard does not accept coating as a means of preventing HISC. Instead, DNV proposes proper material selection and design with respect to stress and strain, even if the material is coated.

DNV proposes a linear stress-strain model for HISC evaluation. The component, e.g., a pipe or a valve, shall be modeled and the stress calculated according to the linear elastic finite element analysis (FEA) stress method. FEA is a computerized method for predicting how a component will react to forces and stresses and/or other physical effects like flow or heat. The stresses

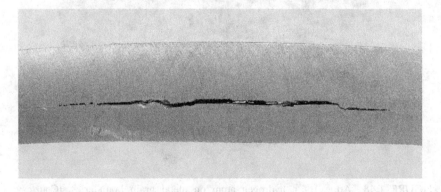

FIGURE 1.49 Hydrogen-induced brittle crack in piping. (Courtesy: Shutterstock.)

Dent in pipeline

FIGURE 1.50   FEA on a pipeline. (Courtesy: Shutterstock.)

and forces, such as axial stress, torsion, bending, etc., are applied to the component, which is divided into small meshes. Figure 1.50 illustrates the result of FEA on a pipeline. The criticality of the load on each mesh area can be recognized from the color of the mesh. Dark blue meshes represent the lowest value of loads. Light blue meshes show the value of forces that are higher than the forces applied to light blue meshes. Red meshes indicate a dent in the pipeline and represent the highest level of loads, i.e., critical loads that would lead to material failure. A dent in a pipeline is defined as an inward movement and plastic deformation that changes the pipe cross-section in the dented area. Yellow and orange meshes typically have stress values between those of the red and light blue meshes.

Linear elastic stress behavior is based on three principles; the first is that stress is always proportional to strain. The second assumption is that the material will be deformed by increasing the loads forever; the last rule is that by releasing the loads, the deformation of the material disappears and the material returns to its initial state. DNV proposes linearizing or converting stress over thickness into two types: membrane stress and bending stress. Peak stress can be disregarded in HISC. Membrane stress is the average stress across the thickness of a component. Bending stress is variable across the thickness of the component and is made of compression and tensile stresses that are applied to the longitudinal axis of the component. Figure 1.51 illustrates the distribution of membrane stress (δ_m), bending stress (δ_b) and bending plus membrane stress (δ_{m+b}) across the thickness of a component.

There are limitations with δ_m and δ_{m+b} in accordance with Equations 1.9 and 1.10:

δ_m limitation

$$\delta_m < \alpha_m \cdot \gamma_{\text{HISC}} \quad \text{SMYS} \tag{1.9}$$

δ_{m+b} limitation

$$\delta_{m+b} < \alpha_{m+b} \cdot \gamma_{\text{HISC}} \quad \text{SMYS} \tag{1.10}$$

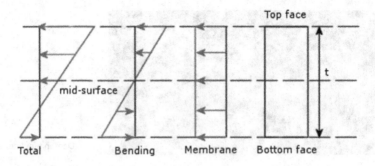

FIGURE 1.51 Distribution of membrane, bending and bending plus membrane stress across the thickness of a component. (Photograph by author.)

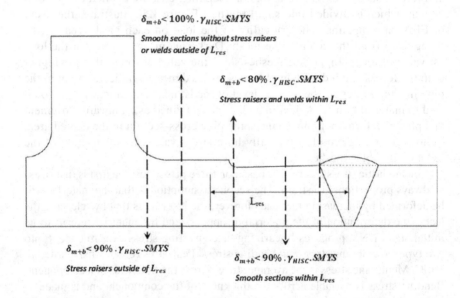

FIGURE 1.52 Values of SMYS factor for membrane plus bending stress from DNV.

Where:

δ_m: Membrane stress (Psi, Pascal);

δ_{m+b}: Membrane plus bending stress (Psi, Pascal);

α_m: Allowable SMYS factor for membrane stress equal to 0.8 (dimensionless);

α_{m+b}: Allowable SMYS factor for membrane plus bending stress equal to 0.8 or 0.9 or 1 depending on the location. The allowable SMYS factor for membrane plus bending stress on weld toes and stress risers within L_{Res} is equal to 0.8. The allowable SMYS factor for membrane and bending stress for stress riser areas outside the L_{Res} is 0.9. The factor is equal to 0.9 and 1 for smooth areas within L_{Res} and outside L_{Res}, respectively. Figure 1.52 illustrates different values of allowable SMYS factor for membrane plus bending stress as per the DNV standard.

TABLE 1.3
Strain Limits as per the DNV Standard

Location in the Component	Strain Limit	
	Within L_{res} from Weld	Outside L_{res} from Weld (%)
Outside 5% WT from any surface	Min. $[0.3\%;0.5\%-\varepsilon_{res}]$	0.3
Within 5% WT from surface	$1.00\%-\varepsilon_{res}$	1.00
	$0.6\%-\varepsilon_{res}$	0.6

γ_{HISC}: HISC material quality factor; equal to 1 or 100% for fine-grain austenitic and 0.85 or 85% for coarse-grain austenitic materials.

The estimation of residual strain ε_{res} is equal to 0.15% at the weld toe (WT) strand 0.25% in the area of L_{Res}. The allowable strain or strain limits for different areas are provided in Table 1.3 as per the DNV standard.

1.5 CONCLUSION

Marine environment is harsh and corrosion and can create various types of corrosion for industrial piping, valves and actuators. Some of these corrosion mechanisms like pitting, chloride stress cracking corrosion (CLSCC) and CUI are explained in this chapter. External coating application is one of the approaches to mitigate these types of corrosion.

1.6 QUESTIONS AND ANSWERS

1. Identify the correct sentence regarding marine environment corrosion.
   A. In the immersion zone, facilities and components are completely submerged in the sea; the immersion zone is the most corrosive zone of the offshore environment.
   B. Seawater may contain corrosive compounds such as chloride, oxygen, carbon dioxide and hydrogen sulfide.
   C. There are three zones in the marine environment: topside, tidal and seawater.
   D. Pitting and CLSCC mechanisms are the only types of external marine corrosion.
   Answer: Option A is not correct; the splash and high tidal zones are the most corrosive zones in the offshore environment, since structures in these zones frequently alternate between being above and below the water due to waves and tidal fluctuations. Corrosion in the splash and high tidal zones is more severe and complex due to the mixture of corrosion and erosion by waves, ice, debris and the corrosive atmospheric marine environment itself, even at times of low tide and minimal waves. Option B is correct; seawater may contain corrosive compounds such as

chloride, oxygen, carbon dioxide and hydrogen sulfide. Option C is not complete because there are four marine environment zones; the splash zone, located between the tidal zone and the atmospheric zone, is missing. Option D is not complete either, as other types of external corrosion exist in the marine environment, such as general corrosion, CUI, galvanic corrosion and HISC.

2. Which statements are wrong in relation to the offshore environment and its zones?

   A. The corrosion rate in the splash zone is typically lower than in the immersion zone.

   B. The corrosion rate in the splash zone can be exacerbated by the erosion effect due to seawater waves.

   C. UV sunlight reduces the corrosion rate and has no effect on protective coatings.

   D. Valve components are typically installed in the splash zone.

   Answer: Option A is not correct because the splash and high tidal zones typically have the highest corrosion rate. Option B is correct; the seawater waves and splash can increase the erosion rate and increase corrosion as a result. Option C is totally incorrect since UV light from the sun both increases the corrosion rate of steel and damages its coating. Option D is not correct, since valves are either installed in the immersion or atmospheric zone. Thus, A, C and D are wrong.

3. Which sentence is correct about a corrosion or electrolytic cell?

   A. An anion is attracted to a cathode and a cation is attracted to an anode.

   B. The flow of electrons is from anode to cathode.

   C. Corrosion occurs in the cathode and the anode is protected.

   D. In an electrolytic cell, the anode is negative and the cathode is positive.

   Answer: Option A is not correct because an anode loses electrons during *oxidation* and becomes positive, and a cathode receives electrons and becomes negative during *reduction* in an electrolytic cell, so the positive anode attracts the anion, which is negative, and the negative cathode attracts the cation, which is positive. Option B is correct; the flow of electrons is from anode to cathode. Option C is wrong because corrosion occurs in the anode and protection occurs in the cathode. Option D is not correct either, as the anode is positive and the cathode is negative in an electrolytic cell.

4. Identify the correct sentences regarding crevice and pitting corrosion types.

   A. One of the main differences between pitting and crevice corrosion is related to the location and the geometry required for corrosion to occur. Unlike pitting corrosion, crevice corrosion takes place in crevices.

   B. Crevice corrosion can only occur on flange faces with serrated finishing.

C. Pitting corrosion occurs in the offshore environment due to the presence of chloride in the environment. It can be prevented by selecting
a proper material.

D. Pitting corrosion is less dangerous than uniform corrosion since it does
not cover the whole surface of metal and occurs in localized areas.

Answer: Option A is correct since crevice corrosion refers to a
localized attack on the metal surface at or close to gaps or crevices
between two jointing surfaces. But pitting corrosion does not necessarily occur in gaps and crevices. Option B is not correct because
crevice corrosion can happen on flange faces with RTJ face finishing in which a groove is located where the gasket ring sits. Option C
is correct; pitting corrosion in the offshore environment is related to
the presence of chloride and can be prevented by selecting a proper
material. Option D is not correct; pitting corrosion is more dangerous
than uniform corrosion because it does not cause corrosion damage
uniformly, so it is more difficult to detect, predict and design against
compared to a uniform corrosion mechanism. Thus, both options A
and C are correct.

5. Which of the following sentences are correct regarding material selection to prevent pitting in the offshore sector?

A. An industrial valve in duplex material in the offshore environment is
operated within a temperature range between a minimum of 120°C
and a maximum of 150°C. No coating is typically required in such a
temperature range.

B. Stainless steel tubes in austenitic stainless steel 316 material can be
upgraded to 6MO to prevent pitting.

C. 6MO has a lower PREN and lower chloride corrosion resistance
compared to super duplex.

D. Both material and coating selection are approaches to mitigating pitting corrosion.

Answer: Option A is wrong since duplex can only be used without
coating at a maximum operating temperature of 100°C, whereas the
temperature, in this case, is above 100°C. Option B is correct; 6MO
tubing as a super austenitic stainless steel can be used as an alternative
to austenitic stainless steel 316 to prevent pitting and CLSCC. It is good
to know that duplex and super duplex tubes could be used instead of
stainless steel 316 in some cases for corrosion prevention. Option C is
not correct, as 6MO has a higher PREN compared to duplex and super
duplex. Option D is correct. Therefore, both options B and D are correct.

6. Which parameters affect the intensity of galvanic corrosion?

A. The surface areas of the anode and cathode; as an example, large
anodic areas that are coupled with small cathodic areas produce
very little galvanic corrosion.

B. The effect of distance, meaning that dissimilar metals in close proximity cause greater galvanic attack.

    C. Higher potential difference between the anode and cathode intensifies galvanic corrosion.

    D. All options are correct.

    Answer: All options are correct, so option D is the correct answer.

7. Identify the correct sentences about CUI.

    A. CUI could be converted to CLSCC in certain materials under specific conditions.

    B. CUI corrosivity is less than corrosion in the topside zone.

    C. Low points, piping joints and bends are at greater risk of CUI compared to straight pipes without any connections.

    D. Coating is the only way to prevent CUI.

    Answer: Option A is correct because the presence of water under insulation could be combined with a source of chloride either from the offshore environment or from the insulation, and with a source of stress and cause CLSCC in a vulnerable material such as austenitic stainless steel 316. Option B is not correct, as the rate of CUI could be more than 20 times higher compared to corrosion in the topside zone. Option C is correct. Option D is not correct, since another approach that can be performed to prevent CUI is to avoid ingress of water under insulation. This can be accomplished by selecting suitable insulation material, properly applying the insulation around the insulated components and protecting the insulation from rain.

8. HISC is highly possible in which of the scenarios mentioned below?

    A. Low-alloy steel bolts for subsea valves that are electroplated with zinc-nickel and subject to cathodic protection.

    B. Subsea piping in 25Cr super duplex that is protected against external corrosion by both coating and cathodic protection.

    C. The stem of a subsea valve in Inconel 725 that is not exposed to seawater.

    D. Welding between a duplex pipe and flange on a topside offshore industry platform.

    Answer: Option A is a scenario that can cause HISC corrosion in the subsea environment because low-alloy steel is a relatively hard material vulnerable to HISC, and the component is bolting that is under constant stress. In addition, using cathodic protection, in this case, can provide the hydrogen required for HISC attack. Option B cannot be considered a scenario with a high risk of HISC corrosion because the piping is coated. A coating without any defect prevents HISC corrosion and hydrogen attack to the metal. Option C does not address a scenario with HISC attack risk, since the stem material is not exposed to seawater even though Inconel 725, which is a hard nickel alloy, is at high risk of HISC. The scenario presented in option D would not cause HISC because it is topside and not subsea.

9. Which material microstructure increases the risk of HISC attack?

    A. Austenitic material structure

B. Fine grains
C. Metal grain flow perpendicular to the source of stress
D. Ferritic material structure

   Answer: Options A and B reduce the chance of HISC attack. Thus, the correct answers are options C and D, which increase the risk of HISC attack.

10. FEA has been applied on an axial check valve in super duplex material. The maximum membrane stress is calculated equal to 140 Mega Pascal (Mpa). The maximum membrane plus bending stress is calculated equal to 180 Mpa. Assuming the SMYS equal to 550 Mpa, the material has coarse grain and the maximum stresses are at the WT, which sentence is correct?
   A. There is no HISC risk in the valve.
   B. The membrane and membrane plus bending stresses cannot cause HISC corrosion.
   C. The value of the HISC material factor is equal to 0.8.
   D. The SMYS factor should be considered equal to 85%.

   Answer: There is no information about strain values in this example, so it is not possible to conclude on the possibility of HISC occurrence; thus, option A is not correct. The value of HISC material for coarse grain is 0.85 and not 0.8, so option C is not correct. The SMYS factor for membrane stress is always equal to 0.8 or 80%, and the SMYS factor for membrane and bending stress at the WT is equal to 0.8 or 80% as per Figure 1.52. Therefore, option D is not correct. The limits for the membrane and membrane plus bending are calculated through Equations 1.11 and 1.12.

Membrane stress calculation

$$\left(\delta_m = 140\,\mathrm{Mpa}\right) < \alpha_m \cdot \gamma_{\mathrm{HISC}}\ \ \mathrm{SMYS} = 0.8 * 0.85 * 550\,\mathrm{Mpa} = 374 \quad (1.11)$$

Membrane plus bending stress calculation

$$\left(\delta_{m+b} = 180\,\mathrm{Mpa}\right) < \alpha_{m+b} \cdot \gamma_{\mathrm{HISC}}\ \ \mathrm{SMYS} = 0.8 * 0.85 * 550\,\mathrm{Mpa} = 374 (1.12)$$

Since both membrane stress and membrane plus bending do not exceed the limits, these two stresses cannot cause HISC failure and option B is correct.

BIBLIOGRAPHY

1. Abioye, T.E. et al. (2015). Laser cladding of Inconel 625 wire for corrosion protection. *Journal of Materials Processing Technology*, Vol. 217, pp. 232–240, Elsevier.
2. American Petroleum Institute (API) 6D (2014). *Specification for Pipeline and Piping Valves*, 24th edition. API, Washington, DC.
3. American Petroleum Institute (API) 20E (2017). *Alloy and Carbon Steel Bolting for Use in the Petroleum and Natural Gas Industries*, 2nd edition. API, Washington, DC.

4. American Society of Mechanical Engineers (ASME) Boiler and Pressure Vessel Code (BPVC) IX (2019). Welding and brazing qualification. New York.
5. Bai, Y. & Bai, Q. (2012). *Subsea Engineering Handbook*, 1st edition. Elsevier, Atlanta, GA.
6. Barke, H.D. et al. (2009). Redox reactions. In: *Misconceptions in Chemistry*. Springer, Berlin, Heidelberg. https://doi.org/10.1007/978-3-540-70989-3_9.
7. Chandler, K.A. (2014). *Marine and Offshore Corrosion*. Marine Engineering Series. Elsevier Science, Amsterdam, Netherlands.
8. Danaee, H.R., et al. (2013). Influence of ultraviolet light irradiation on the corrosion behavior of carbon steel AISI 1015. *Metals and Materials International*, Vol. 19, pp. 217–224, Springer.
9. Det Norske veritas (DNV) RP-F-112 (2008). *Design of Duplex Stainless Steel Subsea Equipment Exposed to Cathodic Protection*. Hovik, Oslo area.
10. Frayne, C. (2010). *Sheir's Corrosion*. Elsevier, Amsterdam, Netherlands, Vol. 4, pp. 2930–2970.
11. Freulon, A., Trinh, A.T., Lacaze, J., Malard, B., & Vu, K.O. (2020). Effect of cast iron microstructure on adherence of an epoxy protection. *International Journal of Cast Metals Research*, pp. 165–170. ISSN 1364-0461.
12. International Organization of Standardization (ISO) 15156 (2015). Petroleum and natural gas industries—Materials for use in H_2S containing environments in oil and gas production. Geneva, Switzerland.
13. Isensee, K. et al. (2018). *Declining Oxygen in the World's Ocean and Coastal Waters*. Intergovernmental Oceanographic Commission of UNESCO, Paris, France.
14. Ivanov, H. (2016). Corrosion protection systems in offshore structures. The University of Akron.
15. Kharisov, B. (2018). *Direct Synthesis of Metal Complexes*. Elsevier Science, Oxford.
16. Makhlouf, A.S.H., et al. (2016). *Handbook of Materials Failure Analysis with Case Studies from the Oil and Gas Industry*, 1st edition. Elsevier, Oxford.
17. National Association of Corrosion Engineers (NACE) (2016). Impact. International measures of prevention, application, and economics of corrosion technology studies. Houston, TX.
18. NORSOK M-001 (2002). *Materials Selection*, 3rd edition. Lysaker, Norway.
19. NORSOK M-501 (2012). *Surface Preparation and Protective Coating*, 6th edition. Lysaker, Norway.
20. Okeremi, A., et al. (2009). Preventing pitting and crevice corrosion of offshore stainless steel tubing. *World Oil Magazine*, pp. 73–80.
21. Raja, V.S. & Shoji, T. (2011). *Stress Corrosion Cracking: Theory and Practice*. Woodhead Publishing, New Delhi.
22. Rorvik, G. et al. (2014). Fasteners in subsea applications: End users experiences and requirements. *International Conference of Offshore Mechanics and Artic Engineering*. American Society of Mechanical Engineers, Paper # OMAE2014-24520.
23. Sotoodeh, K. (2018). Valve failures, analysis and solutions. *Valve World Magazine*, Vol. 23, No. 11, pp. 48–52.
24. Sotoodeh, K. (2020). Optimized material selection for subsea valves to prevent failure and improve reliability. Springer, Berlin, Germany, pp. 1–10.
25. Sotoodeh, K. (2021). *A Practical Guide to Piping and Valves for the Oil and Gas Industry*. Gulf Publishing Professional, an imprint of Elsevier, Austin, TX.
26. Sotoodeh, K. (2021). HISC analysis for valves in the subsea oil and gas industry. *Journal of Safety in Extreme Environments*, Springer. doi: 10.1007/s42797-021-00030-4.

27. Speight, J.G. (2014). *Oil and Gas Corrosion Prevention: From Surface Facilities to Refineries*. Elsevier Science, Amsterdam, Netherlands.
28. Winnik, S. (2016). *Corrosion under Insulation (CUI) Guidelines*, 2nd edition. European Federation of Corrosion. Elsevier, Oxford.

2 Surface Preparation

2.1 INTRODUCTION

Surface preparation is known as the first phase of metal treatment and is an essential action that must be performed before applying coating to a metal surface. All coatings have a specific lifetime design and will fail eventually, but improper surface preparation will result in coating failure at an earlier stage, which is called premature failure. Surface preparation is the most important factor in the performance of any coating system in terms of coating durability, permeability and adhesion. Surface preparation is defined as the treatment of a metal surface before the application of any coating; it can be divided into two major categories: surface cleaning and surface profile or roughness preparation. Without proper surface preparation, even a coating with the highest adhesion will not be durable or cling well. In fact, surface preparation is more important than applying the coating itself. The 95% of coating failures result first from poor surface preparation and second from poor application. It is interesting to note that 85% of these failures occur within the first and second year of coating life during operation. The presence of even a small amount of surface contamination, such as oil or grease, can damage the coating and prevent its proper adhesion to the substrate. The ISO 8501 standard, preparation of steel substrates cleanliness before application of paints and related products, is referred to in the NORSOK M-501 standard for surface preparation. The principal factors that affect clean surface preparation are as follows: removal of rust and mill scale; removal of surface contaminants, including salts, dusts, oils and grease; and finally the surface profile of the substrate. There are three levels of surface preparation: pre-blast preparation, blast cleaning and final surface conditioning. The next section explains the various initial surface conditions and surface contaminants.

2.2 INITIAL SURFACE CONDITIONS

Rust grades are defined according to the Swedish Corrosion Institute in four grades: A, B, C and D, which are illustrated in Figure 2.1. Rust grades are not used in NORSOK M–501 but are mentioned in the ISO 8501-1 standard. Rust grades address the initial surface condition of a metal before applying the surface preparation. Rust grades are defined in the ISO 8501-1 standard as follows:

Grade A: Steel surface is covered completely with adherent mill scale and little if any rust.
Grade B: Steel surface has begun to rust, and the mill scale has begun to flake.

DOI: 10.1201/9781003255918-2

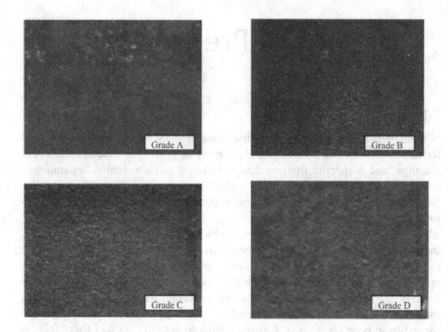

FIGURE 2.1   Rust grades as per ISO 8501-1. (Photograph by author.)

Grade C: Steel surface on which the mill scale has rusted away or from which it can be scraped, but with little pitting visible to the naked eye.

Grade D: Steel surface on which the mill scale has rusted away and on which considerable pitting is visible to the naked eye.

The initial surface of a metal in the form of a pipe or steel structure in preparation for being coated is normally rust grade A or B as per ISO 8501-1. Rust grades C and D are pitted, so these two rust grades should be avoided as much as possible; they would need to undergo surface preparation by sandblasting prior to coating, since it would be very difficult to remove the pitting corrosion products from the metal surface.

2.3   SURFACE CONTAMINANTS

Grease and oil prevent coating from adhering to the surface. Oil and grease are sometimes used on manufactured components for easier assembly. Thus, the first and most important surface treatment is to ensure the complete removal of grease, oil and any other foreign matter. The cleaning of the metal surface to remove grease and oil should be done before rust and mill scale removal and abrasive cleaning.

Rust is an iron oxide with a reddish-brown color (see Figure 2.2) that is formed by the reaction between iron and oxide in the presence of water and oxygen or air and moisture. Rust may adhere firmly to a metal surface or it may be loose. Rust

FIGURE 2.2   Rust on metal. (Photograph by author.)

represents one of the most serious problems for iron and steel as it can cause corrosion. In addition, rusted areas of the steel cannot provide any passivation protection to the underlying iron. As rust is formed naturally, it is difficult to prevent it entirely. Rust left on a metal gets worse and worse and eventually damages and "eats" the metal. Rust results in metal expansion and provides high-stress concentration on the steel. Rust weakens the metal, which becomes flaky and brittle. In fact, rust is mechanically weak, porous and flaky, so it can peel off along with any coating that is applied over it, and it is sensitive to mechanical impact. The amount of rust on a metal surface initially depends on the length of time that the steel is exposed to the environment. The presence of moisture and water can increase rust corrosion. Exposure to outdoor conditions, especially rainy and humid climates, also increases rust corrosion. This is because the rust is permeable to air and water, so the metal beneath the rust layer will continue to corrode. In the presence of salt from a source such as seawater, the process of rusting or rusting corrosion can accelerate. Exposure to sulfur dioxide and carbon dioxide can also speed up the process of rust corrosion.

Luckily, there are some effective approaches that may be taken to prevent rust formation. The first approach is to use a *rust-resistant alloy*. As an example, stainless steels contain a minimum of 11% chromium, which allows the formation of chromium oxide (Cr_2O_3) as a protective layer on the metal surface and prevents rust formation. In fact, the chromium is combined with oxygen before the iron can combine with oxygen to form iron oxide and rust. Many duplex stainless steel piping and valve materials, which are immune to rusting due to their protective layer formation, are used in the offshore industry. Adding nickel can also prevent rusting. Nickel alloys such as Inconel 625, 718 and X750 are used in the offshore oil and gas industry for piping and valve components. Weathering steel, also known as "COR-TEN" steel, contains up to 21% of an alloying element such as chromium, copper, nickel or phosphate. COR–TEN is not a common material for piping and valves in oil and gas industry including offshore although it is cheaper than stainless steel. Additionally, other metals—such as aluminum—and red metals—such as copper, brass and bronze—do not rust. Brass is an alloy of copper and zinc. Bronze is another type of copper alloy with ~12% of tin and addition of other metals such as aluminum, manganese, nickel and zinc. Aluminum cannot rust because rust is iron oxide and aluminum does not have any iron in its composition. Red metals are also not at risk of rusting, as they contain only a neglectable amount of iron. Although these metals do not rust, they can be corroded. As an example, austenitic stainless steel materials are exposed to the risk of pitting and chloride stress cracking corrosion.

The second approach to avoiding rust formation is related to the *design* of the component or product. Undesirable and inappropriate locations, such as cavities and crevices, must be avoided. Welded joints are less exposed to rust formation compared to bolted joints. Figure 2.3 illustrates rust and corrosion on flange joints and their connected bolting in steel material.

A flange is a component used to connect piping, valves and rotating equipment, such as pumps; a flange connection provides easy access to the piping inside for different purposes such as cleaning, maintenance or inspection, which can be accomplished by unfastening the bolts and nuts.

The third approach to preventing rust is to apply *hot-dip galvanizing* (HDG) or zinc electroplating. Applying zinc on a metal surface by means of HDG or zinc

FIGURE 2.3 Corrosion and rust of flange joints and connected bolting. (Courtesy: Shutterstock.)

FIGURE 2.4 Corroded bolts. (Photograph by author.)

electroplating protects the metal surface, as zinc acts as a sacrificial anode and protects the metal. However, applying HDG or zinc electroplating is not always successful in preventing rust and corrosion. It is common in the piping and valve industry to apply HDG on bolts and nuts in carbon and low-alloy steel bolting materials. However, it has been experienced offshore that the HDG is removed from the bolts and nuts after some years of operation; the unprotected carbon or low-alloy steel bolts and nuts are then subject to rust and corrosion as illustrated in Figure 2.4. Thus, 25Cr super duplex bolting has been used recently instead of HDG carbon and low-alloy steel bolts in some Norwegian offshore projects to prevent rust and corrosion.

Bluing is another useful technique that can be used for small steel items to provide protection against rust. The reason why this technique is called "bluing" comes from the blue-black color on the surface of the metal after using this method. Bluing is performed by submerging the steel in a solution of potassium nitrate, sodium hydroxide and water. *Coating* is another means of preventing rust.

The first danger associated with rust is that it can expand and corrode the steel substrate. In addition, rust contains oxygen, moisture and *soluble salt*. In general, "salt" refers to a soluble corrosion product that can sometimes be seen by means of a magnifying glass. However, in general, visual examination is not a reliable way to detect salt on a metal surface. Thus, a number of chemical tests have been developed for the purpose of salt identification. *Soluble salts* can remain on a surface where they have been deposited even after abrasive cleaning of the surface, such as sandblasting. Soluble salts can increase moisture permeation through the coating and increase the corrosion rate under the coating film. Chloride, sulfate and nitrate are the most common types of salts in the coating industry; of these, chloride salt is the most aggressive. Soluble salts can break down coating and can accelerate the corrosion rate of metal by becoming deeply embedded in the iron corrosion product. When old steels are corroded by pitting corrosion, they are more likely to have salts such as ferrous sulfate and iron chloride embedded in the pitted areas.

The ISO 8502 standard is used to assess the cleanliness of surfaces in general. ISO 8502-2 and ISO 8502-5 provide testing methods to assess chloride on a cleaned metal surface before coating application. ISO 8502-6 is used to extract soluble salt for analysis, and 8502-9 can be used to measure the soluble salt on a metal surface. ISO 8502-8 is used to assess the moisture on a metal surface. More information about extraction of the salt from metal surface and analysis is provided in Chapter 4.

White rust is another term that is related to metal surface contamination and preparation. As explained above, galvanizing or zinc coating is a means of providing metal surface protection by adding a layer of zinc. However, zinc can form its own kind of rust, called white rust. White rust is a white, chalky material that can form on the surface of a zinc-galvanized layer. White rust forms on zinc coating when it is exposed to oxygen and hydrogen. The combination of zinc with hydrogen and oxygen will form zinc hydroxide or "white rust," as opposed to normal rust, which is an iron oxide. The contact of the zinc protective layer with water and moisture can form white rust. It is important to know that white rust can be formed easily on a freshly applied zinc or galvanized layer. That is because the new zinc layer has not had a chance to form a stable oxide layer, so hydrogen and oxygen bonds have a chance to react with the zinc and form white rust. White rust is known as a common problem for galvanized steels. The main effect of white rust is to make the galvanized layer useless. Zinc oxide is stable and can provide a very good adhesion to a metal surface, but zinc hydroxide can easily flake off from a metal surface.

It is important to remove the rust from a metal surface before coating implementation, not only to ensure appropriate coating adhesion and implementation but also to save the steel. The rust repair or removal action depends on the stage of rust formation. In general, there are three stages of rust formation: stage 0, stage 1 and stage 2. Stage 0 is an ideal metal surface with regard to rust formation, meaning that no rust is present on the metal surface. Therefore, there is no repair is required at this stage. Since there is no rust on the metal at this stage, no sign of pitting can be seen, and if the metal surface is coated, no sign of coating defect due to the presence of rust can be seen. Although there is no need to repair the surface, some preventive actions can be taken to avoid rust formation by keeping the surface as clean as possible and keeping it away from corrosive substances like water, salt, moisture, etc. Stage 1 is when a rust deposition is found on the surface of the metal. However, the metal surface at this stage is not pitted and has not lost its smoothness. Painting over a rust deposit on a metal surface at stage 1 causes paint defects. At stage 2, rust is intensified and begins to attack the metal surface and cause corrosion. Rust can be removed with a stiff wire brush at stage 2. Thus, *regular maintenance* is advised to stop rust formation and stop the progress of rust from occurring at stage 2.

Mill scale or *scale* is known as another substrate surface contamination. It is formed as a result of hot rolling or heating processes when a metal is heated up to a temperature as high as 1,000°C. When the metal cools down, the surface reacts with the oxygen in the atmosphere and produces mill scale, which is an iron oxide product. Mill scale has a blue-gray color and covers the surface of a metal. It is important to know that mill scale is not a uniform layer and is not well attached to the steel surface. In fact, observing mill scale with a naked eye, it looks like a brittle layer made of small powder particles and chips in a layer of ~1 mm. Mill scale is used initially as a rich source of iron for the production of steel. Mill scale is less reactive and more noble than the steel underneath. Thus, it has a cathodic effect on the steel, such that if it remains on the metal surface or is insufficiently removed from the metal surface, it can initiate and accelerate the galvanic corrosion rate. Corrosion between the steel surface and mill scale causes the mill scale to peel off along with any coating applied on top of it. Mill scale can be removed by means of different approaches, such as flame cleaning, pickling or abrasive blast cleaning, which are explained in more detail later in this chapter.

All contaminants, such as oil, grease, weld flux, dust, rust and chemicals, should be removed during the *pre-blast cleaning* stage. The presence of even a small amount of surface contamination can impair the coating's adhesion to the metal surface. Chemical contaminants that are not visible, such as chloride and sulfates, attract moisture and cause premature failure of the coating. Suitable solvents or degreasing agents should be used to remove the oil, grease and other contaminants. In addition, the surface should be checked for any possible defect, which should be repaired. The importance of surface cleaning and preparation is not limited to cleaning, and the first time the coating is applied. Surface cleaning of the metal surface is also essential during maintenance coating. Coatings that are used for the protection of facilities, components and structures may need maintenance coating after some years of operation.

Dirt and dust on a metal surface prevent the formation of a smooth and uniform film and reduce the adhesion of the coating to the substrate. Blast cleaning, which is explained later in this chapter, produces large quantities of dust and debris that should be removed from the abraded surface after blasting and before the final surface coating. There are different methods of dirt and dust cleaning in the industry, such as mechanical brushing, air blowers, sweep blasting or vacuum cleaning. The challenge is that the effectiveness of applying these cleaning techniques may not be visible. The recommended way to check whether the dust level on the metal surface after cleaning is acceptable is to place a tape on the metal surface. The method of using tape to assess dust presence on a metal surface is described in ISO 8502-3. ISO 8502 addresses the preparation of steel substrates before the application of paints and related products, and part 3 of this standard covers the assessment of dust on steel surfaces that have been prepared for painting (pressure-sensitive tape method). This method is suitable for determining the quantity of dust after blast cleaning of rust grades A, B or C, and the removal of the dust with vacuum cleaning or by blowing with compressed air. Blast cleaning and rust grades are explained in more detail later in this chapter. The pressure-sensitive tape method cannot be used on rust grade D, since the highly corroded areas in rust grade D cannot provide a good adhesion surface for the tape. A 25 mm wide and ~150 mm long piece of tape is placed on the blast cleaned surface and attached thoroughly to the metal surface by rubbing with a thumb or using a soft roller. The tape is then removed and placed on a displaying board of a contrasting color in relation to the dust. The background could be glass, or black or white tiles. Both the quantity and the size of the dust particles are assessed and rated from 0 to 5. Using a magnifying glass is recommended in order to see the dust particle sizes. Figure 2.5 illustrates dust assessment on a metal surface after blast cleaning with a tape as per ISO 8501-3 requirements; the dustier tape at the bottom of the figure indicates a high quantity of dust on the surface, so the metal surface is rejected with regard to dust cleanliness.

To summarize, cleanliness includes the following requirements: no *oil and grease* should be present on the metal surface. Oil and grease should be removed prior to the removal of rusts and mill scale. The presence of even a small amount of oil and grease can reduce coating adhesion and physically damage the coating. Failure to remove the contamination before blast cleaning results in distribution of the oil and grease on the metal surface and contamination of the abrasives. *Rust* is another surface contaminant that causes corrosion and poor adhesion and can result in paint bubbling or blistering. Painting defects, including blistering, are explained in Chapter 4. *Soluble salts* on the metal surface should be removed, as they cause corrosion as well as paint blistering or bubbling. *Mill scale*, which is an iron oxide product similar to rust, is formed in high-temperature conditions, and it is more noble than steel, so it can cause galvanic corrosion if it is not removed from the substrate. In addition, mill scale can cause poor adhesion of the coating to the metal surface. *Dust* is another contaminant that prevents

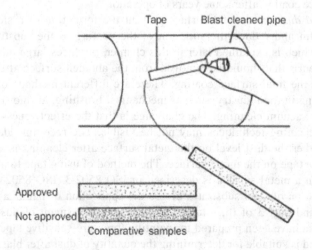

FIGURE 2.5 Dust assessment on a metal surface after blast cleaning with a tape as per ISO 8501-3.

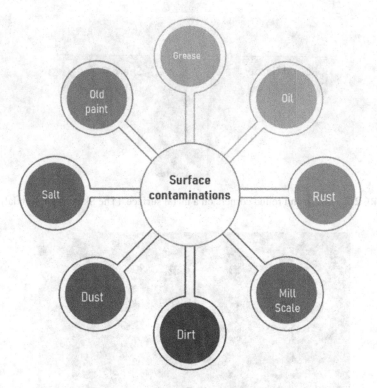

FIGURE 2.6 Metal surface contaminants in a chart. (Photograph by author.)

paint adhesion to the substrate. Figure 2.6 summarizes all of the metal surface contaminants in a chart.

2.4 STEELWORK

The initial work before surface cleaning and blasting as part of surface preparation is called *steelwork*. Steelwork is known as a very important part of surface preparation. All sharp edges should be rounded and smoothed by grinding to a minimum radius of 2 mm since coating adhesion is reduced in sharp areas. Gas-cut edges are sharp and should be ground down before applying the primer. Figure 2.7 illustrates a sharp corner on the right, and a corner that has been rounded by grinding on the left side. But what is the problem with coating sharp edges? Coating a sharp point or area does not provide sufficient coating thickness due to surface tension.

Weld flux or slag, which is a part of the electrode used during welding, provides shielding gas but can increase contamination on the metal surface. All welding spatters (see Figure 2.8), which are droplets of molten metallic material that are scattered and splashed during the welding process, should be ground off. Metal surface defects such as lamination should be removed by grinding during the steelwork activity. Rough manual welds should be ground off and removed. In

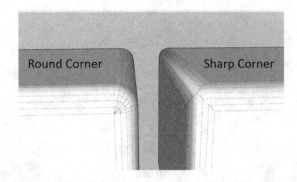

FIGURE 2.7   Sharp and round corners on a metal surface. (The SketchUp Essential)

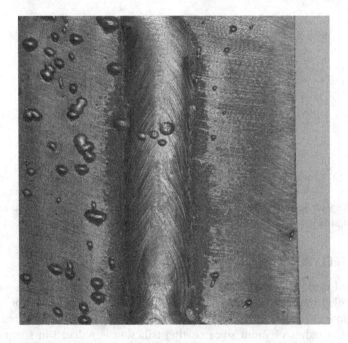

FIGURE 2.8   Welding spatter. (Courtesy: Shutterstock.)

order to avoid contamination and damage to the coating, the steelwork should be performed in the welding shop after welding is completed, and not in the coating shop.

ISO 8501-3, which addresses the surface preparation of metal before coating application, provides a guideline for the preparation of grades of welds, edges and other areas with surface imperfections. Three grades of steelwork preparation are provided in ISO 8501-3: P1, *light penetration*, is defined as the minimum preparation necessary before coating application. P2, *through preparation*, is a grade in which most imperfections have been remediated. P3, *very thorough preparation*,

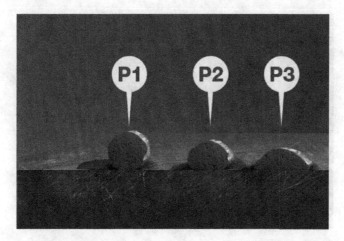

FIGURE 2.9   Welding spatter shapes as per ISO 8501-3. (Courtesy: Graco.)

means that the steel surface is free from significant surface imperfection. Six types of weld imperfection are mentioned in the standard: welding spatter, weld ripple, welding slag, undercut, weld porosity and end craters. Welding spatters are shown in Figures 2.8 and 2.9. P1 surface preparation with regard to welding spatter refers to the surface as it is, without any preparation. P2 preparation grade for weld spatter means that the surface should be free from all completely rounded welding spatters that are loosely adhering to the metal surface. Welding spatter, as shown in Figure 2.9 as shape P1, is considered the loose type of spatter that is loosely attached to the surface and should be removed during P2 preparation. P3 surface preparation means that the metal surface should be free from all welding spatters except for those identified with the bubble labeled "P3" in Figure 2.9 that do not have undercutting.

The second aspect of surface preparation with regard to welding is related to weld ripple or profile, as illustrated in Figure 2.10. Surface preparation P1 means keeping the metal surface as it is welded and shown in the figure. P2 surface preparation indicates a surface that is prepared by removing irregular and sharp edges. P3 surface preparation means that the surface should be fully prepared and completely smooth.

The third surface contamination due to welding is welding slag. Welding slag, also known simply as slag, is material produced as a byproduct of certain arc welding processes. The source of slag, called slag inclusion, is a flux in consumable electrodes used in the arc welding process. Flux covers the core wire, which is in the core of the welding electrode or consumable. It is typically made of silicate and carbonate materials and is melted during the welding process to make a shielding gas and provide a safeguard for the welding pool. The welding pool refers to the welding area where the base material has reached its melting point and is ready for fusion with the welding electrode. Figure 2.11 illustrates welding slag on a welding surface. Welding slag must be removed from the

FIGURE 2.10   Welding ripple or profile. (Courtesy: Graco.)

FIGURE 2.11   Welding slag on a metal surface after welding. (Courtesy: Graco.)

surface; it is not allowed in any of the three surface preparation grades (P1, P2 or P3).

Undercut is the fourth type of welding defect that can occur during steelwork. Undercut is a type of welding defect in which the weld reduces the cross-section thickness of the base metal. One reason for this type of defect is excessive current, which causes the edge of the joint to melt and drain into the weld. Figure 2.12 illustrates undercut welding defects in some areas of a weld. Surface preparation P1 with regard to undercut means that the surface remains as it is without any preparation. P2 means that the surface should be free from sharp undercuts, and P3 means that the surface shall be free from deep or jagged undercuts.

Undercut

FIGURE 2.12 Welding undercut. (Courtesy: Graco.)

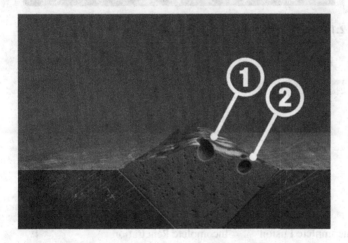

FIGURE 2.13 Welding porosity types as visible and invisible as per ISO 8501-3. (Courtesy: Graco.)

The other welding defect that should be noted is welding porosity (see Figure 2.13), meaning the presence of cavities in the welds caused by freezing the gas from the weld pool. The definitions of different surface preparations with regard to the ISO 8501-3 standard and weld porosity are as follows: P1 grade means that the weld porosities will remain as they are without any surface treatment. P2 grade surface preparation means that the surface pores should be open and visible, as illustrated in number 1 in Figure 2.13. Having open and visible pores allows for the penetration of paint. The highest grade of steelwork preparation is P3, in which the metal surface should be free from porosity, including visible pores.

The last welding defect and imperfection mentioned in ISO 8501-3 with regard to surface preparation are end craters. A crater refers to the end of the weld or weld stop. If the weld crater end fusion leg dimension falls below the requirement, it is considered a weld defect (see Figure 2.14). In that case, since the crater does

FIGURE 2.14   Welding crater defect. (Courtesy: Graco.)

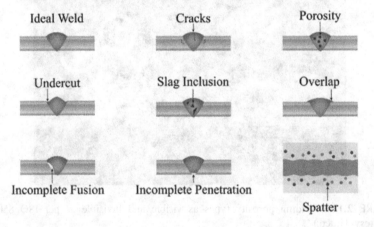

FIGURE 2.15   Some important weld defects. (Courtesy: Shutterstock.)

not provide the required fusion, it is not considered a part of the acceptable leg of the welding. As illustrated in the figure, the end part of the weld is not completely filled in due to poor welding technique during welding stop.

P1 surface preparation in regard to weld craters means no preparation; keep the crater as it is. P2 surface preparation permits end craters but without any sharp edges. P3 surface preparation means that the surface of the steel should be free from welding craters. The recommendation of the present author is to always prepare the steelwork according to a P3 level of surface preparation according to the ISO 8501-3 standard. In addition, ISO 12944-3, the standard for corrosion protection of steel structures by protective paint systems—part 3—design considerations, states that welds should be free from imperfections such as undercut, spatter, etc. Some of the most important weld defects are summarized in Figure 2.15.

When it comes to steel preparation related to edges, four types of edges are described in ISO 8501-3: rolled edges, edges with punching and shearing and thermally cut edges; these are all illustrated in Figure 2.16. These types of edges can be created during fabrication by means of punching, shearing and drilling tools.

When it comes to surface preparation related to rolled edges, P1 and P2 surface preparation grades indicate that the surface should be kept as it is without any preparation. Thus, both P1 and P2 are not acceptable surface preparation for coating. Alternatively, P3 surface preparation requires that the edges be rounded with a radius not <2 mm. P1 surface preparation for punching and shearing on the metal surface means that no part of the edge should be sharp. P2 surface preparation grade with regard to punching and shearing indicates that they should be smooth, and P3 requires punching and shearing edges to be rounded with a radius not <2 mm. For thermally cut edges, P1 indicates that the surface should be free from slag and loose scales. P2 means that no part of the edge should have an irregular profile, and P3 indicates that cut faces shall be removed and cut edges should be rounded with a radius not <2 mm.

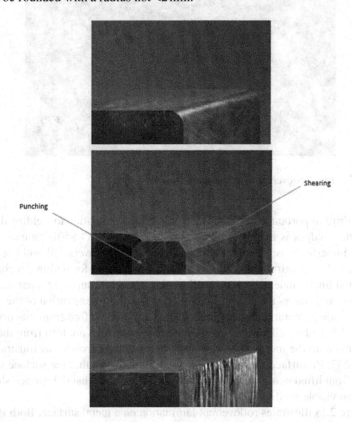

FIGURE 2.16   Rolled edge (a), punching and shearing (b) and thermally cut edge (c). (Courtesy: Graco.)

Shelling or sliver

FIGURE 2.17   Shelling or slivering on a metal surface. (Courtesy: Graco.)

FIGURE 2.18   Rollover/cut lamination. (Courtesy: Graco.)

The third important factor in surface preparation in relation to welding defects and surface edges is associated with surface defects. ISO 8501-3 mentions six surface imperfections: pitting and craters, shelling, rollovers/roll and cut laminations, rolled-in extraneous matter, grooves and gouges formed by mechanical action and finally indentations and roll marks. P1 and P2 surface preparation pits allow pits and craters that are sufficiently open to allow penetration of the welds. But P3 grade preparation means that the surface shall be free from pits and craters. Shelling, also called slivering, refers to segments that are torn from the steel and rolled onto the metal surface, mainly because of corrosion, as illustrated in Figure 2.17. P1 surface preparation for shelling indicates that the surface should be free from lifted material. However, P2 and P3 indicate that the surface shall be free from visible shelling.

Figure 2.18 illustrates rollover/cut lamination on a metal surface. Both P2 and P3 surface preparation grades indicate the surface of the metal should be free from rollovers and laminations. Rollover is a type of operation on the metal done

FIGURE 2.19   Rolled-in extraneous matter. (Courtesy: Graco.)

FIGURE 2.20   Grooves and gouges on a metal surface. (Courtesy: Graco.)

to change the shape of the metal, reduce its thickness or both. If the roller moving over the metal surface cuts some parts of the metal by mistake, it will cause ununiform metal thickness and metal slices.

Figure 2.19 illustrates rolled-in extraneous matter on a metal surface. This type of defect happens during fabrication when foreign matter is caught under the mechanical roller and embedded in the steel surface. All three grades (P1, P2 and P3) indicate that the surface of the metal should be free from rolled-in extraneous matter.

Groves and gouges (see Figure 2.20) are typically formed by mechanical action due to mishandling. Grooves and gouges should be left as they are as per P1 surface preparation. P2 surface preparation indicates that the radius of the grooves

FIGURE 2.21   Indentations and roll marks. (Courtesy: Graco.)

shall not be <2 mm. P3 surface preparation indicates that the surface should be free from grooves and gouges.

Indentations and roll marks can be made on a metal surface by a heavily loaded roller or by mechanical manipulation. Indentations and roll marks are illustrated in Figure 2.21. P1 surface preparation means leaving the defect as it is. P2 surface preparation indicates that indentations and roll marks shall be smooth, and P3 requires the surface to be completely free from indentations and roll marks.

To summarize, steelwork commonly involves the following stages and steps before surface cleaning and applying primer on the substrate:

- All sharp edges must be rounded by grinding to a minimum radius of 2 mm. Other types of edges, such as gas-cut edges, should be rounded off and ground down before applying paint.
- All welding slag shall be ground off. In addition, other rough manual welding defects should be removed.
- Undercut as a weld defect should be repaired completely. Undercutting in the welding should be repaired before primer application.
- All surface defects, such as shelling/slivering and lamination, should be removed.

The figures below show some examples of unacceptable metal surfaces that should be avoided for coating. Spot welding is not acceptable as a means of preparing a metal surface for coating, as it can lead to crack and crevice corrosion as illustrated in Figure 2.22. Figure 2.23 illustrates porosity in a fillet weld on a steel structure that should have been removed but was painted over mistakenly.

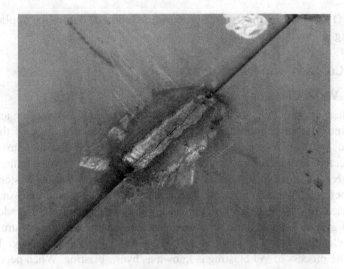

FIGURE 2.22 Spot welding and crack (welding defect). (Courtesy: Shutterstock.)

FIGURE 2.23 Steel structure that was coated despite porosity in the weld. (Courtesy: Shutterstock.)

2.5 SURFACE CLEANING METHODS

There are three stages of metal preparation: *pre-blast cleaning, blast surface cleaning* and *final surface preparation*. Surface cleaning is divided into three or four steps: cleaning with liquid or gas, mechanical and/or blast cleaning and cleaning and dusting before coating implementation. The selection of cleaning method depends on different parameters, such as substrate material, geometry of the components that should be cleaned, degree of surface cleanliness and profile and the location where the surface preparation takes place. ISO 8504 addresses surface preparation methods before the application of coating.

2.5.1 Cleaning with Liquid or Gas

2.5.1.1 Water Cleaning

Fresh water cleaning (WC) is a good method for removing salts, fouling and any loose paint or other contaminants from the substrate. The pressure of the WC should be 7–10 bar. As a general rule, the cleaning process becomes faster and more effective with increasing water pressure. Additionally, increasing the water pressure results in using less water. ISO 8501-4:2006, preparation of steel substances before application of paints and related products—visual assessment of surface cleanliness—Part 4: Initial surface conditions, preparation grades and flash rust grades in connection with water jetting, mentions high-pressure water jet (WJ) cleaning with a pressure of more than 700 bar equal to 70 Mega Pascal (Mpa). The process of WJ blasting is known as hydro-blasting. When necessary, ultra-high-pressure water jetting with a pressure of 25,000 psi equal to ~1,724 bar can be used for metal surface cleaning. Ultra-high-pressure water jetting is popular because it can remove soluble salts from a steel surface.

The Society for Protective Coatings (SSPC) visual guide VIS 4/National Association of Corrosion Engineers (NACE) VIS 7 provides two definitions for WC. Water cleaning is a term used for WC with a pressure of <70 Mpa equal to 700 bar or 10,000 psi. Low-pressure water cleaning (LPWC) is defined as cleaning performed at a pressure <7 Mpa equal to 70 bar or 1,000 psi. High-pressure water cleaning (HPWC) is defined as WC performed at a pressure from 7 to 70 Mpa (70–700 bar or 1,000–10,000 psi). As mentioned above, WC with a pressure above 70 Mpa or 700 bar is defined as water jet (WJ) cleaning. Using a WC technique with a pressure above 170 Mpa equal to 25,000 psi or almost 1,700 bar is called ultra-high-pressure water jetting as per the SSPC and NACE definition. However, ISO 8501-4 defines WJ cleaning as having a pressure above 1,400 bar equal to 21,000 psi. Figure 2.24 illustrates the removal of old, red paint from a ship by means of water jet cleaning before sandblasting and applying the new coating. Table 2.1 provides a summary of WC definitions, pressure ranges and flow rates.

Three grades of surface appearance after WC are defined in ISO 8501-4: Wa 1, Wa 2 and Wa 2½. Wa 1, *light high-pressure water jetting*, means that, when viewed without magnification, the surface shall be free from visible oil and grease, loose or defective paint, loose rust and other foreign matter. Any residual contamination shall be randomly dispersed and firmly adherent. Wa 2, *thorough high-pressure water jetting*, means that, when viewed without magnification, the surface shall be free from visible oil, grease and dirt, most of the rust, previous paint coatings and other foreign matter. Any residual contamination shall be randomly dispersed and can consist of firmly adherent coatings, firmly adherent foreign matter and stains of previously existent rust. Wa 2½, *very thorough high-pressure water jetting*, means that, when viewed without magnification, the

FIGURE 2.24 Water jet cleaning of a ship. (Courtesy: Shutterstock.)

TABLE 2.1
Water Cleaning Definitions, Pressure and Flow Rate Values

Definition	Pressure (bar)	Pressure (psi)	Flow Rate (L/min)
Low-pressure water cleaning	70 maximum	1,000 maximum	
High-pressure water cleaning	70–700	1000–10,000	50–90
High-pressure hydro-blasting	700–1,700	10,000–25,000	25–50
Ultra-high-pressure hydro-blasting	>1,700	>25,000	12–25

surface shall be free from all visible rust, oil, grease, dirt, previous paint coatings and, except for slight traces, all other foreign matter. Discoloration of the surface can be present in places where the original coating was not intact, and any grey or brown/black discoloration observed on pitted and corroded steel cannot be removed by further water jetting.

ISO 8501-4 addresses WJ cleaning as a method of surface preparation before coating application. In addition, the SSPC visual guide VIS 4/NACE VIS 7 has developed a guide and reference photographs for steel surfaces prepared by water jetting. It is important to know that WJ cleaning is covered by SSPC–SP–12 /NACE 5. Four grades of water jet cleaning are defined in SSPC VIS 4/NACE VIS 7: WJ1, WJ2, WJ3 and WJ4. The definition of each WJ grade is provided as follows: *WJ1, clean to bare substrate*: The surface shall be cleaned to a finish that, when viewed without magnification, is free from all visible rust, dirt, previous coatings, mill scale and foreign matters. *WJ2, very thorough or substantial cleaning:* The surface shall be cleaned to a finish that, when viewed without magnification, is free from all visible oil, grease, dirt and rust except for randomly

FIGURE 2.25   Initial metal surface in rust grade C. (Photograph by author.)

dispersed stains of rust, tightly adherent thin coating and other tightly adherent foreign matter. The stain or tightly adherent matter is limited to a maximum of 5% of the surface in WJ2. *WJ3, thorough cleaning:* The surface shall be cleaned to a finish that, when viewed without magnification, is free from all visible oil, grease, dirt and rust except for randomly dispersed stains of rust, tightly adherent thin coating and other tightly adherent foreign matter. The stain or tightly adherent matter is limited to a maximum of 33% of the surface in WJ3 surface preparation. *WJ4, light cleaning:* The surface shall be cleaned to a finish that, when viewed without magnification, is free of all visible oil, grease, dirt, dust, loose mill scale, loose rust and loose coating. All residual compounds that are tightly adherent should remain on the surface. Figure 2.25 illustrates a metal surface in rust grade C before applying WJ cleaning. Rust grades, which could be A, B, C or D, are explained above. Figures 2.26–2.29 illustrate the substrate surface after applying WJ4, WJ3, WJ2 and WJ1, respectively.

WJ cleaning can be effective in removing water-soluble surface contaminants that cannot be removed by dry abrasive method. In fact, WJ cleaning can help remove grease and oil, rust, welding spatter and existing coating and lining from a metal surface. Cleaning with water can be implemented for initial metal surface preparation or for removal of loose adherent paints. The efficiency and impact of the WC can be increased by other approaches rather than increasing the pressure, such as adding detergent or increasing the temperature of the water. Adding detergents to the water can easily remove oil, grease, soluble salts and dusts, but attention should be paid to rinsing the metal surface with clean, fresh water afterward. The temperature of the steel can rise during the WJ process because compressing the water to reach the high jetting pressure increases the water temperature. This temperature increase can be substantial; it accelerates drying time and increases the water's effectiveness in removing rust.

The application of WJ cleaning to remove the remaining paints and coating from a metal surface rather than using abrasive blasting has the advantage of

FIGURE 2.26 Metal surface after applying WJ4. (Photograph by author.)

FIGURE 2.27 Metal surface after applying WJ3. (Photograph by author.)

avoiding the health hazards to which personnel are exposed during sandblasting; moreover, it does not leave any abrasive disposal and removes salts from the metal surface more easily than blast cleaning. Since no abrasive is typically used, the surface will not obtain a new roughness profile unless some abrasives are added to the water. It is possible to add some abrasive to a WJ treatment for surface cleaning to achieve a rougher surface profile, but of course adding abrasive to the jet water for cleaning makes the cleaning process more expensive. Back to the advantages of WJ cleaning, no dust is produced by WJ cleaning, unlike blast cleaning. The coat of WJ is lower compared to grit abrasive blasting. The noise level is lower in WJ cleaning and no dust is produced, in contrast to sandblasting.

Wet abrasive blasting means using liquid as the third element for blast cleaning. The first two elements that are implemented for standard or dry blast cleaning

FIGURE 2.28   Metal surface after applying WJ2. (Photograph by author.)

FIGURE 2.29   Metal surface after applying WJ1. (Photograph by author.)

are abrasive media and compressed air. Wet abrasive cleaning has some advantages, such as better removal of salts from the metal surface and no dust creation compared to standard sandblasting. The disadvantage of wet abrasive cleaning is that it causes *flash rust*. Flash rust can form within just a few minutes to a few hours after cleaning completion.

2.5.1.2   Steam Cleaning

Steam cleaning is a cleaning process that involves using low-pressure steam to remove soluble substances from a metal surface. It can be used to remove surface contaminants such as oil, grease, salt and dirt. Detergent may be added to the steam for better cleaning; in such a case, the substrate should be washed and cleaned afterward with fresh water. The definition used for steam cleaning in this subsection is provided according to ISO 12944-4. ISO 12944 is titled Paints and varnishes—Corrosion protection of steel structures by protective paint systems; part 4 of this standard refers to types of surfaces and surface preparation.

2.5.1.3 Emulsion Cleaning

Emulsion cleaning is performed to remove oil, grease, salt, dirt and other contaminants. Water rinsing should be done after emulsion cleaning. The WC should be fresh and may be hot or cold. Emulsion cleaning is mentioned in ISO 12944-4 as a cleaning technique and metal surface preparation.

2.5.1.4 Alkaline Cleaning

Alkaline cleaning is performed to remove oil, grease, salt, dirt and other contaminants. Water rinsing should be done after alkaline cleaning. Alkaline cleaning is mentioned in ISO 12944-4 as a cleaning technique and metal surface preparation.

2.5.1.5 Organic Solvent Cleaning

Solvent cleaning is a cleaning process used to remove unwanted grease, oil, dirt, salt and other contaminants, such as paint, from a material surface. There are different types of solvents on the market, and care must be taken in order to select the most suitable one. Solvent cleaning is frequently used to remove the grease from a metal surface after a machining operation or to remove damaged or incorrectly applied coating. Organic solvent cleaning is mentioned in ISO 12944-4 as a cleaning technique and metal surface preparation.

2.5.1.6 Chemical Cleaning

Steam cleaning, emulsion cleaning, alkaline cleaning and organic solvent cleaning are all methods of chemical cleaning. Chemical or solvent cleaning is covered by SSPC–SP 1. SSPC–SP1 is a requirement for surface preparation that removes contaminants from steel surfaces. Solvent cleaning is the primary method of removing visible dirt, grease, oil, soil, drawing compounds and similar organic compounds from steel surfaces. Generally, mill scale and other inorganic compounds cannot be removed by any type of solvent cleaning, with the exception of acid pickling. SP1 is an important requirement before surface preparation by blast cleaning, such as SSPC–SP 10/NACE 2, near-white blast cleaning, or SSPC–SP 5/NACE 5, white metal blast cleaning. Blast cleaning should not be performed on a surface with grease and oil, as grease and oil could contaminate the metal shot or grit. Sandblasting is not an effective grease and oil cleaning process, which is why solvent or chemical cleaning methods are performed before sandblasting. In general, tightly adherent contaminants, stains, streaks and shadows remain on the metal surface 100% after solvent or chemical cleaning. Chemical cleaning is not effective even in removing loosely adherent materials to the metal surface. Figure 2.30 illustrates the degreasing solvent cleaning of a rusted flange surface to remove oil, grease and rust.

Pickling and passivation are types of chemical cleaning, more precisely acid cleaning, that are common for stainless steel. The surface of stainless steel should be clean, smooth and faultless. The cleaning of stainless steel is important,

FIGURE 2.30 Chemical/solvent cleaning of a flange face to remove rust, oil and grease. (Courtesy: Shutterstock.)

because it increases the corrosion resistance of the material, prevents contamination and achieves the desired appearance. A combination of degreasing, pickling and passivation is required to achieve the required cleanliness of a surface of stainless steel. A smooth surface is important to prevent cracking and flaking and to prevent the adherence of surface contaminants to the stainless steel. Pickling and passivation are not effective in removing grease from metal surfaces, so degreasing may also be required. Pickling and passivation are typically combined with blast or mechanical cleaning. Pickling is the most common chemical means of removing mill scale (iron oxide contamination) and low chromium layers. Different types of acids are involved in pickling, such as nitric acid (HNO_3), hydrofluoric acid (HF) and sometimes sulfuric acid (H_2SO_4). Hydrochloric acid is prohibited due to the effect of chloride on stainless steel and pitting corrosion. Acid pickling is covered by SSPC–SP 8 as well as ISO 8504-4. It is good to know that the effectiveness of the acids is increased by increasing the temperature. Pickling in the form of spraying rather than immersion is suitable for large surfaces. Passivation is another type of chemical acid treatment that is common for stainless steel to remove contaminations such as iron contamination and promote the formation of a passive film of chromium oxide (Cr_2O_3). Nitric acid is a very common type of acid for passivation, which uses less aggressive chemical cleaning compared to pickling. Passivation should be applied after mechanical cleaning. Mechanical cleaning can produce iron contamination, which can be removed by passivation. Iron contamination is different from rust. In fact, any steel or iron item that comes in contact with stainless steel, such as the steel grit or shot used in sandblasting, are sources of iron contamination in stainless steel. If iron remains on the stainless steel for a while, it can mix with oxygen and create rust.

Electropolishing is an electrochemical process in which the workpiece is submerged in an electropolishing electrolyte and connected to a positive terminal (cathode). As a result of the electropolishing process, the atoms are removed from the workpiece and converted to ions. Typically, oxidation takes place in the anode, and reduction occurs in the cathode by producing hydrogen. Although any

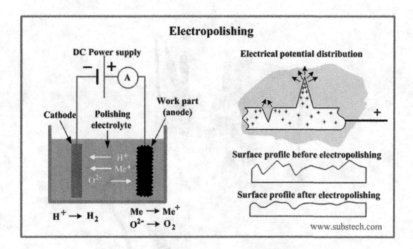

FIGURE 2.31   Electropolishing process. (Courtesy: SubsTech.)

metal can be electropolished, the most commonly polished metals are austenitic and martensitic stainless steels, copper, aluminum, titanium, nickel and copper alloys. The electropolishing process (see Figure 2.31) reduces surface roughness and makes the surface brighter. The other advantages of electropolishing are improving the corrosion resistance of the material, relieving stress and improving the material's resistance against fatigue. Electropolishing does not involve abrasive materials and has a higher Health, Safety and Environment (HSE) compared to blast cleaning and provides smoother, lower-friction surfaces. In addition, electropolishing can be used as a steelwork method alternative to mechanical polishing to remove welding defects or polishing rough surfaces such as cut and rolled edges. The disadvantage of electropolishing is that very rough surface defects cannot be removed by this technique.

The opposite process of electropolishing is *electroplating*, in which a coating such as copper, silver or gold is deposited on a metal that is immersed in an electrolyte solution. Figure 2.32 illustrates the silver plating of a spoon with a silver strip immersed in an electrolyte.

2.5.2   Blast Cleaning

Coating on machine surfaces is subject to early degradation due to poor adhesion. Some past studies have shown that machined surfaces provide less contact surface than blasted samples. *Blast cleaning* includes the usage of silica, sand, steel shot or steel grit and involves the impinging action of these abrasive materials (see Figure 2.33). In the industrial coating industry, the metal surface is not only blasted to remove surface contaminations but also to generate a surface profile conducive to coating adhesion.

Abrasive blasting, also called *sandblasting* or *shot blasting*, is a generic term for the high-speed application of abrasive materials against the surface of a metal

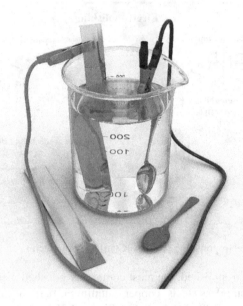

FIGURE 2.32 Electroplating process to deposit silver on a spoon. (Courtesy: Shutterstock.)

FIGURE 2.33 Steel shoot and steel grit for abrasive or blast cleaning. (Photograph by author.)

to smooth a rough surface, roughen a smooth surface, shape a surface or remove surface contaminants and metal dust for better coating adhesion. Sandblasting can eliminate substances such as rust, paint, oil, etc. ISO 8504-2 addresses the abrasive blast cleaning method of surface preparation. Decreasing the density of peaks or areas of roughness on a metal surface in general results in better adhesion between the coating and metal surface. Abrasive blasting or sandblasting can be used to remove the old paint or corrosive products from the metal surface. The important point is that the abrasives themselves should be clean and free from

Before   After

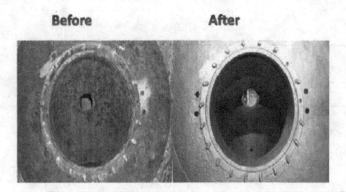

FIGURE 2.34 Metal specimen surface before and after sandblasting. (Photograph by author.)

contamination. Figure 2.34 illustrates the surface of a metal specimen before and after sandblasting.

An air compressor is used as a power source that directs the air with a pressure of at least 100 psi to blast the sand fast enough to clean the metal to the desired degree. The air for abrasive blasting or sandblasting should be free of dust, mists and gases, and must meet the relevant air quality requirements. The other method of blasting is *shot blasting*; the key difference between sandblasting and shot blasting is that a centrifugal or mechanical force is used for shot blasting. A device similar to a spinning wheel is used to centrifugally accelerate the sand or shot and blast it against the surface. In fact, we can say that shot blasting is more aggressive than sandblasting. As explained later in this chapter, sandblasting can cause serious health problems such as respiratory disease for personnel who work with sands.

One important factor in surface blast cleaning is the surface profile obtained during blast cleaning. The surface profile, also simply called "profile" or "profile roughness," is an important key for coating adhesion. The metal surface profile depends on various parameters, such as the type of abrasive(s) used, the air pressure and the blasting technique. Surface profile is a unique factor that is independent from the standard and degree of cleanliness. While too low a surface profile does not provide a good enough surface for coating application, too high a surface profile results in uneven coating with sharp peaks on the surface that can cause premature coating failure. Roughness itself is a series of microscopic peaks and valleys across a surface. Figure 2.35 illustrates the roughness profile of a metal surface before blast cleaning. The first parameter illustrated in the figure is the distance between the tallest peak and the deepest valley in the surface R_t. The second parameter in the figure is roughness average (R_a), which is defined as the average distance to an imaginary center line that can be drawn between the peaks and valleys. Surface roughness is calculated by measuring R_a, the average of the surface heights and depths across the surface. The ASME B46.1 standard defines R_a as the arithmetic average of absolute values of the profile height deviations

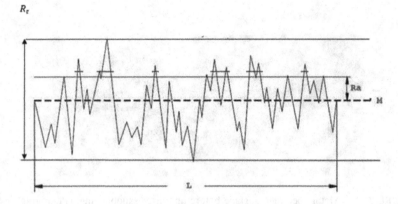

FIGURE 2.35   Surface profile indicating parameters R_t and R_a. (Photograph by author.)

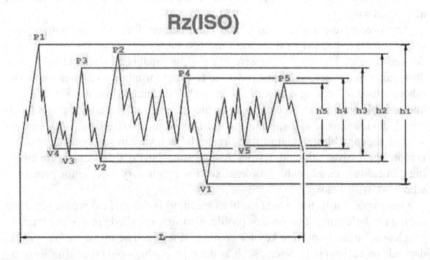

FIGURE 2.36   Surface profile. (Photograph by author.)

from the mean line recorded within the evaluation length. Figure 2.36 illustrates R_z, which is the average value of the distance from the peaks to the valleys, which is equal to the blasting profile. As shown in Figure 2.36, the top five values of the peaks are connected to the top five values of the valleys, which result in distances of h_1, h_2, h_3, h_4 and h_5. The average of h_1, h_2, h_3, h_4 and h_5 is known as R_z, which could be four to six times the R_a.

The main advantages of sandblasting are that it is easy and popular, requires less machinery and equipment and provides an excellent cleaning surface. Sandblasting is very easy to apply and does not require a skilled worker. The only machinery used in the sandblasting process is a compressor. But the disadvantages of sandblasting are mainly associated with its health hazard. The sand is made from silica in many cases, and inhaling silica puts humans at risk of respiratory illnesses such as silicosis. Silica dust is also known as a cause of lung cancer. The United States Occupational Safety and Health Administration (OHSA) has studied the effects of sandblasting on American personnel working with silica particles and inhaling its dust. OHSA did not ban the usage of silica sand for abrasive cleaning but did create some rules and safety regulations for using and working with sands. The other challenge associated with using sands is that they typically contain contaminants and moisture, which should be removed before usage.

As discussed earlier in this chapter, there are two methods of abrasive blasting: dry blasting and wet blasting. Dry blasting involves two elements: blasting components and compressed air. Wet blasting includes a third element, which is a liquid. Table 2.2 summarizes the advantages and disadvantages of dry and wet blasting.

Although wet blasting is messier compared to dry blasting, it is worth applying to remove dust and to avoid the health hazards associated with dry blasting.

2.5.2.1 Blast Cleaning Equipment

Blasting can take place in a portable blast pot, blast cabinet or blast room. Mobile or portable blast pots or equipment (see Figure 2.37) contain dry abrasive blast systems that are typically powered by a compressor. The compressor can provide a large volume of air to a single blast pot or to multiple blast pots. Blast pots are pressurized with compressed air and filled with abrasive materials. A blast cabinet (see Figure 2.38) is a type of closed-loop system that allows the operator to recycle the abrasives. A blast cabinet consists of four different parts: containment or casing, an abrasive blasting system, an abrasive recycling system and a dust collection system. Sandblasting can also be applied in a blast room, which is much larger than either a portable blast pot or blast cabinet, as illustrated in Figure 2.39. The larger size of the blast room provides the possibility of blasting larger components and equipment. Manual cleaning and coating inside a blast room can be performed by an operator, and sometimes the coating process is

---

TABLE 2.2

Wet and Dry Sandblasting

Advantages	Disadvantages
Wet Blasting	
Better removal of salts	Flash rust development on the surface is possible after cleaning
No dust creation	
Dry Blasting	
Surface remains dry	Not a good method for removing salt and oil
No possibility of rust formation after cleaning	Creates rust

FIGURE 2.37   Portable blast pot. (Photograph by author.)

FIGURE 2.38   Blast cabinet or abrasive blast chamber machine. (Courtesy: Shutterstock.)

accomplished by using manipulators and robots. The interior of the room should be ventilated, and a dust collector should be supplied as part of the blast room system to keep the room dust-free.

Sandblast cleaning of very large components can take place in an open area, as illustrated in Figure 2.40. It is noticeable that sandblasting a large pipe in an open area

FIGURE 2.39   Sandblasting room. (Courtesy: Shutterstock.)

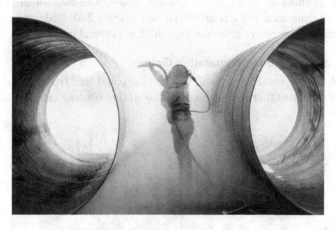

FIGURE 2.40   Sandblasting of a large pipe produces a great amount of dust. (Courtesy: Shutterstock.)

produces a great amount of dust. The chamber or portable blast pot where the sand and air are mixed is not shown in the figure. In the sandblasting process, the compressor provides an air supply to the portable blast pot. The mixture of the sand and air travel through a hose to reach the hand-held nozzle. The high-pressure sand and air finally leave the nozzle and shoot toward the surface of the workpiece.

The production rate of sandblasting depends on three essential parameters: the efficiency of the system and capacity of the compressor and the pressure at the nozzle. As an example, if the compressor produces 100 psi of pressure and the same pressure exists at nozzle, then the system is 100% efficient. However, if the compressor produces 60 psi air pressure rather than 100 psi and there is only 50 psi at the nozzle, the efficiency is 50%. Lower efficiency means that additional time will be required to complete the blast cleaning. Nozzle and hose size also affect the cleaning rate. The design of the compressor, nozzle and blast pot or sandblaster are very important for the efficiency of the sandblasting. The first important consideration is to size the compressor to constantly produce a pressure of 100 psi at the nozzle at the required volume. The second important consideration is related to pressure drop between the compressor and the nozzle. The length, internal diameter and roughness of the hose affect the pressure drop. A smoother internal hose surface will reduce the pressure drop between the compressor and the nozzle. A larger hose also results in less pressure reduction between the compressor and the nozzle. Selection of a proper sandblaster is also essential for the design of the sandblasting system. The last step is to select a nozzle size that is appropriate for the selected size of the compressor. Jotun, a Norwegian multinational chemicals company specializing in coating, proposes a hose three to four times bigger than the nozzle opening or orifice. It is possible to distinguish the smaller size of the blast nozzle compared to the connected hose in Figure 2.41. Reducing the nozzle size increases the pressure at the nozzle, which is a critical parameter.

2.5.2.2   Blast Cleaning Preparation Grades

The *final surface condition* should be clean, dry and free from grease and oil. Dusts and abrasives that are used for blasting during blasting cleaning should be

FIGURE 2.41   Comparing nozzle and hose size. (Courtesy: Shutterstock.)

TABLE 2.3
Blast Cleaning Roughness Profile
Grades as per ISO 8503-1

Blast Cleaning with Grit Abrasives (G)

Fine (G)	25–50 μm
Medium (G)	50–85 μm
Coarse (G)	85–130 μm

Blast Cleaning with Shot Abrasives (S)

Fine (S)	25–35 μm
Medium (S)	35–60 μm
Coarse (S)	60–85 μm

removed. The metal surface cleanliness and roughness are the key parameters for the final surface condition. The surface roughness or profile of the substrate is covered by the ISO 8503-1 standard. There are two types of roughness measurement defined in the ISO 8503-1 standard: blast cleaning profile grade G with abrasive grit, and blast cleaning profile S with abrasive shot. Both types of abrasive blast cleaning (G and S) are divided into three categories of roughness: fine, medium and coarse. Table 2.3 summarizes the abrasive surface roughness values for grit and shot as per the ISO 8503-1 standard. In general, shot abrasive blasting provides less surface roughness compared to the correlated grit blast cleaning with the same category of roughness.

Four standard cleanliness conditions are commonly used according to the ISO 8501-1 and 2 standards, preparation of steel substrates before application of paints and related products: Sa 1 (PSa 1), Sa 2 (PSa 2), Sa 2½ (PSa 2½) and Sa 3 (PSa 3). It is important to specify the required cleanliness standard and roughness condition of the substrate or blast cleaning surface preparation grades in the coating specification. The surface preparation (cleanliness) grades in ISO 8501-1 and ISO 8501-2 are the same, but ISO 8501-2 addresses the preparation grades of surfaces that had been coated previously, and from which the coating has not been removed completely. ISO 8501-1 addresses metal surfaces that are either uncoated or from which all the previous coating has been removed. Both standards contain pictures to illustrate the preparation grades. The important point is that the letter "P" is related to surface preparation according to ISO 8501-2, and indicates that the metal surface retains some coating from a previous application.

Sa 1 is a blast cleaning surface preparation known as a *light blast*, in which rapid blast cleaning is performed to remove loose mill scale, rust and foreign matters. Sa 1 does not provide an acceptable surface preparation for long-lasting painting implementation, which is typically required in offshore environments. The term "sweep blasting" is used to refer to mild or light blasting with relatively low blasting force or low-pressure blast cleaning. Sweep blasting is not covered in any standard, but it can be considered similar to Sa 1, as it provides mild and light blasting. Sweep blasting, as a mild, abrasive blasting, can be used to remove loose paint attached to a metal surface before applying a suitable surface preparation and or applying maintenance coating. Figure 2.42 illustrates a flange connection in a piping system in the offshore environment that had been coated initially. But the coating is damaged and corrosion can be observed on the pipe and flange connection. Sweep blasting can be applied on such a flange and piping surface in order to remove the loose and damaged coating before applying surface preparation and maintenance coating. Figure 2.43 illustrates the details of a metal surface with damaged coating and corrosion before applying sweep blasting.

FIGURE 2.42 Coating damage and corrosion on piping and a flange in the offshore environment. (Courtesy: Shutterstock.)

FIGURE 2.43 Metal surface with loosely adherent coating and corrosion before sweep blasting. (Courtesy: Shutterstock.)

Sweep blasting can be used after abrasive blast cleaning to remove particles, dirt and dust from a metal surface. To sum up, Sa 1 can remove loosely adhering materials 100%, but all tightly adhering materials, stains, streaks and shadows will remain on the surface. Sa 0 means no surface preparation.

Sa 2, known as *thorough* or *commercial* blast cleaning, removes almost all the mill scale, rust, rust scale, paint and foreign matters by means of abrasives. Contaminants that are firmly attached and adhering to the metal surface will not remain on the surface after SA 2 blast cleaning. Commercial blasting is not typically acceptable in the NORSOK M-501 standard, as it does not provide an excellent or perfect metal surface. But we can say that some degree of cleanliness is achieved. This type of surface preparation could be applied prior to coating in non-corrosive environments and atmospheres. After applying abrasive blasting, the surface is cleaned with a vacuum cleaner, dry compressed air or a clean brush. Mill scale, often just called scale, is a type of iron oxide that is formed on the surface of steel during the hot metalworking process. To sum up, all loosely adhering materials, tightly adhering materials and 65%–66% of stains, streaks and shadows are removed from the metal surface after Sa 2 metal surface cleaning. Sa 2 has a grayish color, unlike Sa 1, which has a black and reddish color due to the presence of rust and foreign matters. Figure 2.44 compares Sa 1 and Sa 2 metal surface abrasive cleaning.

Sa 2½ is known as *very thorough blast cleaning* in which almost all mill scales, rust and foreign matter are removed. In fact, all loosely and tightly adhering materials are removed completely as a result of thorough blast cleaning. Sa 2½ removes 85% of stains, streaks and shadows, and provides a near-white metal cleaning surface as illustrated in Figure 2.45. Finally, the surface of the metal is cleaned with a vacuum cleaner, clean dry compressed air or a clean brush. Near-white blast cleaning is typically required and specified for high-performance coating to be used in severe environmental conditions such as offshore oil and gas industry atmosphere.

Sa 3 is known as the highest grade of blast cleaning to "pure metal" or "white metal." This grade of blast cleaning removes all loosely and tightly adhering materials as well as stains, streaks and shadows. Finally, the surface is cleaned with a vacuum cleaner, clean dry compressed air or a clean brush.

FIGURE 2.44 Sa 1 and Sa 2 metal surface comparison after abrasive cleaning. (Photograph by author.)

FIGURE 2.45   Sa 2½ metal surface after abrasive cleaning. (Photograph by author.)

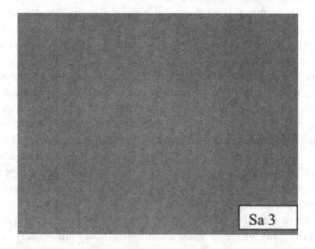

FIGURE 2.46   Sa 3 metal surface. (Photograph by author.)

The metal surface after blasting to Sa 3 is completely and uniformly gray, as illustrated in Figure 2.46. White metal is specified for steel materials that are exposed to high temperature, high pressure and a corrosive environment, and in cases where coating failure can cause catastrophic consequences, such as nuclear reactors.

The Society for Protective Coatings (SSPC) is an American National Standards Institute accredited standards development organization. SSPC develops and publishes widely used industry standards for surface preparation, coating selection, coating application, painting contractor certification and testing. SSPC specifies the following important blast cleaning surface preparation levels:

SSPC–SP 5 (NACE 1): White metal blast cleaning

Completely remove all mill scale, rust, rust scale, previous coating, etc., leaving the surface a uniform gray-white color. This is very high-quality surface cleaning, equivalent to ISO Sa 3, in which all loosely and tightly adherent contaminants, as well as stains, streaks and shadows are removed from the metal surface.

SSPC–SP 6 (NACE 3): Commercial-grade blast cleaning

Completely remove all dirt, rust scale, foreign matter and previous coating, etc., leaving shadows and/or streaks caused by rust stain and mill scale oxides. Random staining shall be limited to no more than 33% of each unit area of surface (a unit of area is defined as 9 square inches). SP 6 is equivalent to ISO Sa 2.

SSPC–SP 7 (NACE 4): Brush-off blast cleaning

Remove rust scale, loose mill scale, loose rust and loose coatings, leaving tightly bonded mill scale, rust and previous coatings. This is an ideal method for removing oxides and/or loose and peeling coatings from galvanized metal. Results are comparable to those achieved by thorough chipping, scraping and wire brushing. SP7 is equivalent to ISO Sa 1 and sweep blasting, and it can be selected for surface cleaning as a replacement for SP 2 and 3, hand and power cleaning tools.

SSPC–SP 10 (NACE 2): Near-white metal blast cleaning

Remove all loosely and tightly adherent rust scale, mill scale, previous coating, etc., leaving only light stains from rust, mill scale and small specks of previous coating. Random staining shall be limited to no more than 5% of each area of surface. It can provide higher surface cleaning than ISO Sa 2½ and does not have any ISO equivalent.

SSPC–SP 14 (NACE 8): Industrial blast cleaning

Remove all visible oil, grease, dust and dirt. Traces of tightly adherent mill scale, rust and coating residues are permitted to remain on 10% of each unit area of the surface if they are evenly distributed. The traces of mill scale, rust, and coating shall be considered tightly adhered if they cannot be lifted with a dull putty knife. Shadows, streaks and discoloration caused by stains of rust, stains of mill scale and stains of previously applied coating may be present on the remainder of the surface. Industrial blast cleaning is a degree of cleanliness between SP 7 (Sa 1) and SP 6 (Sa 2) and does not have any ISO equivalent. Figure 2.47 summarizes the essential blast cleaning surface preparation levels based on SPCC and the equivalent ISO SP standard.

The other important consideration is that after applying the surface preparation according to the acceptable standard and the required surface grade and profile, the surface of the metal could deteriorate. Re-rusting can occur on the substrate very quickly in wet environmental conditions. Therefore, it is essential to apply coating on the metal surface as soon as possible and practical after surface preparation, especially when the surface preparation is done with a grinding or power tool. Postponing the coating application on the metal surface and keeping the steel in a corrosive environment can cause re-rusting and additional cost and time of coating.

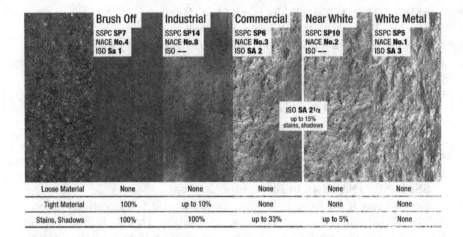

	Brush Off	Industrial	Commercial	Near White	White Metal
	SSPC **SP7** NACE **No.4** ISO **Sa 1**	SSPC **SP14** NACE **No.8** ISO —	SSPC **SP6** NACE **No.3** ISO **SA 2**	SSPC **SP10** NACE **No.2** ISO —	SSPC **SP5** NACE **No.1** ISO **SA 3**
				ISO **SA 2½** up to 15% stains, shadows	
Loose Material	None	None	None	None	None
Tight Material	100%	up to 10%	None	None	None
Stains, Shadows	100%	100%	up to 33%	up to 5%	None

FIGURE 2.47 Blast surface cleaning as per SSPC and equivalent ISO grades. (Courtesy: Graco.)

FIGURE 2.48 Hand wire brush. (Photograph by author.)

2.5.3 Mechanical Cleaning

Mechanical cleaning means using machines such as steel brushes, grinding equipment or processes such as machining. It is important to note that mechanical cleaning cannot provide surface cleanliness as clean as that achieved by blast cleaning. Mechanical cleaning can be divided into two categories: hand tool cleaning and power tool cleaning. Hand tool cleaning is covered by SSPC–SP2, and power tool cleaning is covered by SSPC–SP3. Hand tool cleaning is a method of steel preparation that involves using non-power hand tools. Using a hand wire brush (see Figure 2.48), a coated abrasive disk, a non-woven abrasive disk and sandpaper are methods of hand tool cleaning. Wire brushes can be categorized into hand wire brushes (hand tool) and mechanical wire brushes (power tool). In general, surface cleaning by means of hand tools such as wire brushes and sand-paper is not effective in completely removing rust and mill scale. Power tools can offer better surface preparation compared to manual or hand tools. St 2 and St 3 are "thorough localized hand and power tool cleaning" and "very thorough local-ized hand and power tool cleaning," and can be achieved by hand and power tool cleaning, respectively. St 2 is defined as "thorough scraping (with a hard-metal scraper) and wire brushing, disc-sanding, etc. During the operation, all the loose scale, rust and foreign matter should be removed. Finally, the surface is cleaned with a vacuum cleaner, clean and dry compressed air or a clean brush. It should then have a faint metallic sheen." St 3 is defined as "Extremely thorough scraping and wire brushing, disc sanding, power brushing, etc. Surface preparation as for St 2, but considerably more accurate. After removing the dust, the surface should have a pronounced metallic sheen." Just like Sa grade preparations, the letter "P" is added before St surface preparation grades if the metal surface had been coated previously and the coating has been partially removed (e.g., P St 2 and P St 3).

A wire brush can remove loose rust, mill scale and paint. The remaining con-taminants, which adhere tightly to the metal surface, cannot be removed by wire brush, sandpaper or coated abrasive disk. Figure 2.49 illustrates the removal of loose rust from a substrate by means of a mechanical wire brush. In addition, unlike sandblasting, no surface profile can be achieved by the hand tool clean-ing methods mentioned above. A coated abrasive disk or wheel (illustrated in Figure 2.50) can be used for light to medium grinding and for removing loose paint, mill scale and rust.

Sandpaper is another hand tool, which consists of a sheet of cloth or paper with abrasive materials attached or glued to its surface. Sandpaper, like a wire brush, can be used to remove loose rust, mill scale and paint from a metal surface without making a surface roughness or profile. Figure 2.51 illustrates a hand-held

FIGURE 2.49 Removal of rust from a metal surface with a mechanical wire brush. (Courtesy: Shutterstock.)

FIGURE 2.50 Abrasive wheel or disk. (Courtesy: Shutterstock.)

FIGURE 2.51 Sandpaper used to polish the rough surface of a metal. (Courtesy: Shutterstock.)

sandpaper tool being used to polish the rough surfaces of a metal. An alternative for hand and power tool cleaning, SSPC–SP2 and SP3, respectively, is brush-off grade blasting (SSPC–SP7). SSPC–SP7, which is equivalent to NACE 4 and ISO Sa 1, removes rust scale, loose mill scale, loose rust and loose coatings, leaving tightly bonded mill scale, rust and previous coatings. As explained above, SP 7 is called brush-off or sweep blast cleaning. Sweep blast cleaning removes 100% of the loosely adherent contaminants, but 100% of tightly adherent materials, stains, streaks and shadows remain on the metal surface. This is an ideal method for removing oxides and/or loose and peeling coatings from galvanized metal. The results are comparable to those achieved through chipping, scraping and wire brushing.

Two other SSPC preparation grades with power tool cleaning, SSPC–SP11 and SSPC–SP 15, provide a higher degree of surface cleaning and can be performed as alternatives to sandblast cleaning. SSPC SP11, "Power tool cleaning to bare metal," provides a minimum surface profile of 1 mil. Unlike SSPC–SP 3, which can be used only to remove loose contamination from the substrate, SSPC–SP 11 provides a surface profile free of visible oil, grease, dirt, dust, rust, coating, oxide, mill scale, corrosion products and foreign matters. However, a slight residue of rust and coating is permitted to remain after surface preparation according to SSPC–SP 11. In fact, the power tool cleaning method of SSPC–SP 11 provides a good surface profile for paint adherence. The second difference between SSPC–SP 3 and SP 11 is that no surface profile is required by SP 3. The second power tool standard, SSPC–SP 15, "Commercial grade power tool cleaning," like SSPC–SP 11, asks for the requirement of a 1 mill surface profile and bringing the surface to bare metal, but unlike SSPC–SP 11, SSPC–SP 15 allows random staining to remain on the metal surface.

A grinding machine, also simply called a grinder, is a type of power or machine tool that is used for grinding. The main feature of a grinder is an abrasive wheel that acts as a cutting tool. The abrasive wheel is made of several abrasive grains that cut small chips from the workpiece. Figure 2.52 illustrates a worker cleaning a welding seam on a piece of steel. The abrasive disk or wheel in the grinder, also called a coated abrasive tool, acts like sandpaper but with mechanical power. Figure 2.53 illustrates a worker grinding welded metal spatter from the substrate

FIGURE 2.52   A worker cleans a weld seam with an electrical wheel grinding machine. (Courtesy: Shutterstock.)

FIGURE 2.53   Grinding of welded metal spatter. (Courtesy: Shutterstock.)

FIGURE 2.54   Grinding a sharp edge. (Courtesy: Shutterstock.)

with an electrical (power) wheel grinding machine. The sharp edge of a square cross-section area of metal is being ground and rounded in Figure 2.54. Both hand and power tool cleaning surface preparation methods are covered by the ISO 8504-3 standard.

Comparing blast cleaning and mechanical cleaning, we can say that blast cleaning can provide an ideal substrate compared to mechanical surface cleaning. Manual tools, like hand wire brushes, are not recommended since they provide a very poor surface finish. Power wire brushing is better than manual wire brushing, but it should typically be followed by a polishing job, unlike power tool surface preparation as per SSPC–SP 11 and 15. Power grinding is not as good as blast cleaning, but it could be the best alternative to blast cleaning among the mechanical cleaning choices. The other disadvantage of surface preparation by grinding or power tools could be the necessity of applying the paint as soon as possible to prevent re-rusting and surface contamination of the prepared metal surface.

Table 2.4 summarizes the main SSPC standards and the equivalent NACE and ISO standards. The SSPC VIS standard provides a visual guide and reference photographs for surface cleanliness.

TABLE 2.4
Surface Preparation Standards Comparison

Description	SSPC Standard	NACE Standard	International ISO Standard
Solvent cleaning	SSPC–SP1		
Hand tool cleaning	SSPC–SP2		ISO 8501-1 St2
			ISO 8501-2 PSt2
Power tool cleaning	SSPC–SP3		ISO 8501-1 St3
			ISO 8501-2 PSt3
White metal blast cleaning	SSPC–SP5	NACE 1	ISO 8501-1 Sa 3
			ISO 8501-2 PSa3
Commercial blast cleaning	SSPC–SP6	NACE 3	ISO 8501-1 Sa 2
			ISO 8501-2 PSa 2
Brush-off blast cleaning	SSPC–SP7	NACE 4	ISO 8501-1 Sa 1
			ISO 8501-2 PSa 1
Pickling	SSPC–SP8		
Near-white metal blast cleaning	SSPC–SP10	NACE 2	Almost ISO 8501-1 Sa 2 ½
			Almost ISO 8502-2 PSa 2 ½
Power tool cleaning to bare metal	SSPC–SP11		
Water jetting	SSPC–SP12	NACE 5	
Industrial blast cleaning	SSPC–SP14	NACE 8	
Commercial-grade power tool cleaning	SSPC–SP15		
Brush-off coated and uncoated galvanized steel, stainless steel and non-ferrous metals	SSPC–SP16		
Abrasive blast cleaning	SSPC–VIS 1		
Power and hand tool cleaning	SSPC–VIS 3		
Water jetting	SSPC–VIS 4	NACE–VIS 7	
Wet abrasive blast cleaning	SSPC–VIS 5	NACE–VIS 9	

2.6 CONCLUSION

Surface preparation is known as one of the most important tasks to be performed prior to coating. Poor surface preparation of metal causes premature failure of coating at an early stage. Various surface preparation methods like sandblasting, WJ cleaning, chemical cleaning as well as various metal surface preparation regarding cleanliness and roughness.

2.7 QUESTIONS AND ANSWERS

1. Which sentences are not correct regarding rust and mill scale?
   A. Rust and mill scale are completely similar as both of them are made of iron oxide.
   B. Material protection against rust does not mean that corrosion will not occur on the protected material.
   C. Aluminum does not rust because of the protective layer of aluminum oxide on the metal surface.
   D. Rust can cause corrosion, as rusted areas on the metal surface cannot provide passivation of the areas underneath, and corrosion by rust can be enhanced by seawater and exposure to carbon dioxide.
   Answer: Option A is not correct. There are differences between rust and mill scale; mill scale is an iron oxide formed due to high temperature during the hot rolling process, while rust can be formed by a metal's exposure to a corrosive environment and does not necessarily require high-temperature exposure in order to occur. The other difference between mill scale and rust is color; mill scale is blue-gray in color, while rust is reddish brown. Option B is correct, as some materials such as austenitic stainless steel are immune to rust but can be corroded by certain corrosion mechanisms, such as pitting and chloride stress cracking corrosion. Option C is not correct; the reason why aluminum does not rust is because it does not contain any iron. Option D is also correct, so in conclusion both options B and D are correct.

2. Which statements are correct regarding metal surface contaminants?
   A. Oil and grease could remain on a metal surface during abrasive cleaning since they are removed effectively by abrasive sands.
   B. Mill scale is more noble than metal, so it can cause galvanic corrosion if it is not removed completely from the metal surface.
   C. Soluble salts can create corrosion, lack of paint adhesion and paint blistering, so they should be removed from the substrate before coating is applied. High-pressure WJ cleaning is an effective method for removing salts.
   D. Dirt and dust are unavoidable after wet blast cleaning, so visual inspection is a common way to check whether or not the amount of dust on the metal surface is acceptable.

Answer: Option A is not correct; the first stage of metal surface cleaning before applying abrasive blasting is to remove the oil and grease. If oil and grease are present, sandblasting can spread them on the metal surface and contaminate the abrasives. Thus, the presence of oil and grease on a steel surface during sandblasting causes the failure of the abrasive blasting operation. Option B is correct; mill scale is more noble than metal, and it causes galvanic corrosion in contact with the metal surface if it remains on the substrate partially or completely. Option C is also correct regarding the negative effects of soluble salts in terms of corrosion and paint blistering. In addition, high-pressure WJ cleaning is an effective way to remove salts. In fact, WJ metal surface cleaning is popular because of its effectiveness in removing salts. Option D is not completely correct for two reasons: First, using wet abrasive blasting with liquid leaves no dust after cleaning. Second, although dirt and dust are unavoidable after dry blast cleaning, visual inspection is not a common way to check the dust level on a metal surface. ISO 8502-3 recommends using a tape to assess the cleanliness of the metal surface. Therefore, in conclusion, options B and C are correct statements about metal surface contamination and cleaning.

3. Which options are incorrect regarding WJ surface cleaning?
   A. SSPC VIS 4 and NACE VIS 7 define three grades of WJ cleaning: Wa 1, Wa 2 and Wa 2½.
   B. Any water pressure above 700 bar, which is almost equal to 10,000 psi, is called water jet.
   C. WJ grades that have higher numbers in front provide better cleaning.
   D. ISO 8501-4 provides three grades for WJ cleaning.

   Answer: Option A is not correct, because SSPC VIS 4/NACE VIS 7 specify four grades for water jet cleaning: WJ1, WJ2, WJ3 and WJ4. Option B is not completely correct, since any water pressure starting from 25,000 psi is called "ultra-high-pressure" WJ cleaning as per both NACE and SSPC definitions. Therefore, the WJ pressure can be considered from 700 bar almost equal to 10,000 psi to 1,700 bar almost equal to 25,000 psi. Option C is not correct; lower grade numbers indicate better cleaning compared to higher grade numbers. As an example, WJ1 specifies much cleaner surface preparation compared to WJ4. Option D is correct, since the ISO 8501-4 standard, initial surface conditions, preparation grades and flash rust grades in connection with water jetting, provides three grades: Wa 1, Wa 2 and Wa 2½.

4. Which activity or activities can be performed on a metal surface before sandblasting?
   A. Steelwork
   B. Cleaning the oil and grease with solvent
   C. Salt removal by water
   D. All three options can be carried out before sandblasting

Answer: All the options are correct.

5. Which type of surface preparation is performed on stainless steel mainly to activate the protective chromium oxide layer and remove low chromium containing layers?
   A. Mechanical cleaning
   B. Chemical cleaning
   C. Sandblasting
   D. WJ cleaning

   Answer: Pickling and passivation is a type of chemical cleaning performed on stainless steels to remove low chromium layers and promote a protective layer of chromium oxide. Thus, option B, chemical cleaning, is correct.

6. Which sentence is correct regarding the standard measurements of surface preparation?
   A. Sa 1 is defined as no surface preparation, while Sa 0 provides light blast cleaning when a rapid blast cleaning is performed to remove loose mill scale, rust and foreign matter.
   B. Sa 2 is very thorough blast cleaning in which all the tightly adhering materials, including stains, streaks and shadows, are removed from the metal surface.
   C. Sa 2 blast cleaning is suitable for a long-lasting coating system in a corrosive offshore environment.
   D. The NORSOK M–501 coating standard requires Sa 2½ for all coating systems except coating system 6.

   Answer: Option A is not correct because Sa 0 is defined as no surface preparation. The second part of option A provides a definition for Sa 1 and not Sa 0. Option B is not completely correct, as Sa 2 is called thorough blast cleaning but it cannot remove all streaks and shadows. In fact, Sa 2 removes just 65%–66% of streaks and shadows. Sa 3 is a blast cleaning of white metal, which removes all streaks and shadows. Option C is not correct, since Sa 2, also called commercial blasting, provides an acceptable metal surface for non-corrosive environments, so it is not suitable for offshore use. Option D is the correct answer; NORSOK M-501, which is the coating standard in the Norwegian petroleum industry, requires Sa 2½ for all coating systems except coating system 6.

7. Identify the correct sentences regarding steelwork and the surface preparation of metals as per the ISO 8501-3 standard.
   A. Two types of metal surface imperfections are addressed in ISO 8501-3; one is welding defects, and the other is related to surface defects in general.
   B. All types of edges should be rounded with a radius not less than 2mm; refer to ISO 12944-3.
   C. Grooves and gouges are a type of surface defect as per the ISO 8501-3 standard.

D. Rolled and thermally cut edges are the only two edge types mentioned in ISO 8501-3.

Answer: Option A is not correct since three types of surface imperfections are addressed in ISO 8501-3: welding defects, edges and general surface defects. Option B is correct, since all three types of edges should be prepared according to surface preparation grade P3, meaning that the edges should be rounded with a radius not <2 mm as per ISO 12944-3. Option C is correct; grooves and gouges are a type of surface defect as per ISO 8501-3. Option D is not correct, because additional edge types are defined in ISO 8501-3, namely punching and shearing.

8. Identify the incorrect sentences with regard to surface preparation standards.

A. Surface preparation according to SSPC–SP 5 provides a less clean surface compared to ISO 8501 Sa 3.

B. The ISO 8501-1 surface preparation standard addresses the preparation of metal surfaces with some coating left on the surface.

C. Power tool cleaning should be always performed before sandblasting.

D. American Society for Testing and Materials (ASTM) standards do not address coating and painting at all.

Answer: Option A is wrong since SSPC–SP 5 provides a cleanliness level equal to ISO 8501 Sa 3. Option B is wrong since ISO 8501-1 is a standard for preparing a surface that has not been coated, or from which the coating has been removed completely before surface preparation. ISO 8501-2 is the correct standard for surface preparation when some coating is left on the surface. Option C is not always correct. Power tool cleaning as per SSPC–SP 3, which provides surface preparation equal to St 3, is typically applied before sandblasting. However, power tool cleaning of bare metal and commercial power tool cleaning, according to SSPC–SP 11 and SSPC–SP 15, respectively, provide enough cleanliness and surface adhesion for coating, so the latter-mentioned power tool cleaning can be used as an alternative to blast cleaning. Option D is not correct; some ASTM standards do address coating. Thus, all four options are wrong.

9. Which of the following statements are correct regarding hand and power tool cleaning?

A. Hand and power tool cleaning are categorized as kinds of mechanical cleaning.

B. The ISO 8504-4 standard covers hand and power tool cleaning.

C. SSPC–SP 11 provides cleaner power tool surface preparation compared to SSPC–SP 3 and SSPC–SP 15.-

D. The best SSPC sandblasting replacement for SSPC–SP 2 and 3 is SSPC–SP 5.

Answer: Option A is correct. Option B is not correct, because ISO 8504-3 covers hand and power tool cleaning and ISO 8504-4 is about acid pickling. Option C is correct; SSPC–SP 3 addresses basic surface cleaning with a power tool, which can remove loose contaminants from the metal surface. SSPC–SP 15 does not allow any contaminant, such as paint, mill scale, rust, etc., but does allow random staining on the metal surface. SSPC–SP 11 is like SSPC–SP 15, but no staining is allowed by SSPC–SP 11. Thus, SSPC–SP 11 provides a cleaner surface compared to the other two standards. Option D is not correct. The best replacement for SSPC–SP 2 and 3 (surface preparation by hand and power tool cleaning) is SSPC–SP 7, brush-up blast cleaning. Brush-up blast cleaning is like sweep blasting and can provide a surface cleanliness of Sa 1 as per the ISO 8501-1 standard. However, SSPC–SP 5 provides a white metal blast cleaning surface, which is equivalent to a surface cleaning of Sa 3 as per the ISO 8501-1 standard. The best replacement for SSPC–SP 2 and 3 is SSPC–SP 7 known as brush-off blast cleaning provides a surface cleaning of Sa 1. Thus, options A and C are correct.

10. Which ISO standards are used for assessing the cleanliness of a metal surface with regard to dust, chloride and salts?

A. ISO 8501-1
B. ISO 8501-4
C. ISO 8503-1
D. ISO 8502-2, 3, 5 and 9

Answer: Option A is not correct, as ISO 8501-1 is a standard for *defining* the cleanliness of a metal surface, not assessing its cleanliness. Option B is not correct, as ISO 8501-4 has to do with the WJ surface cleaning method. Option C is not correct either, since ISO 8503-1 is about the surface roughness of a substrate before applying the coating. Option D is the correct answer. ISO 8502-2 and 5 address the chloride assessment of a metal surface; ISO 8502-3 and ISO 8502-9 address the assessment of dust and salt on a metal surface before coating, respectively.

BIBLIOGRAPHY

1. American Society of Mechanical Engineers (ASME) B46.1 (2019). Surface texture (Surface Roughness, Waviness, and Lay). New York.
2. Bahadori, A. (2015). *Essentials of Coating, Painting, and Lining for the Oil, Gas and Petrochemical Industries*. Gulf Professional Publishing, an imprint of Elsevier, Waltham, MA.
3. Finishingsystems (2019). Shot blasting & sandblasting: What's the difference? [online] available at: http//www.finishingsystems.com/blog/shotblasting-sandblasting-difference [access date: 21th March 2021].
4. Graco (2021). [online] available at: https://www.graco.com/ [access date: 11th April 11, 2021].
5. Hagen, C.H.M., et al. (2016). The effect of surface profile on coating adhesion and corrosion resistance. Paper # NACE-2016-7518, Vancouver, British Columbia, Canada.
6. International Organization of Standardization (ISO) 12944-3 (2017). Paint and varnishes – Corrosion protection of steel structures by protective paint systems – Part 3: Design considerations. Geneva, Switzerland.

7. International Organization of Standardization (ISO) 12944-4 (2017). Paint and varnishes – Corrosion protection of steel structures by protective paint systems – Part 4: Types of surface and surface preparation. Geneva, Switzerland.

8. International Organization of Standardization (ISO) 8501 (2007). Preparation of steel substrates before application of paints and related products – Visual assessment of the surface cleanliness – Part 1: Rust grades and preparation grades of uncoated steel substrates and of steel substrates after overall removal of previous coating. Geneva, Switzerland.

9. International Organization of Standardization (ISO) 8501 (2007). Preparation of steel substrates before application of paints and related products – Visual assessment of the surface cleanliness – Part 2: Preparation grades of previously coated steel substrates after localized removal of previous painting. Geneva, Switzerland.

10. International Organization of Standardization (ISO) 8501 (2006). Preparation of steel substrates before application of paints and related products – Visual assessment of the surface cleanliness – Part 3: Preparation grades of welds, edges and other areas with surface imperfections. Geneva, Switzerland.

11. International Organization of Standardization (ISO) 8501 (2020). Preparation of steel substances before application of paints and related products – Visual assessment of the surface cleanliness – Part 4: Initial surface conditions, preparation grades and flash rust grades in connection with water jetting. Geneva, Switzerland.

12. International Organization of Standardization (ISO) 8502 (2017). Preparation of steel substrates before application of paints and related products – Tests for the assessment of surface cleanliness – Part 2: Laboratory determination of chloride on cleaned surfaces. Geneva, Switzerland.

13. International Organization of Standardization (ISO) 8502 (2019). Preparation of steel substrates before application of paints and related products – Tests for the assessment of surface cleanliness – Part 3: Assessment of dust on steel surfaces prepared for painting (pressure–sensitive tape method). Geneva, Switzerland.

14. International Organization of Standardization (ISO) 8502 (1998). Preparation of steel substrates before application of paints and related products – Tests for the assessment of surface cleanliness – Part 5: Measurement of chloride on steel surfaces prepared for painting (ion detection tube method). Geneva, Switzerland.

15. International Organization of Standardization (ISO) 8502 (2006). Preparation of steel substrates before application of paints and related products – Tests for the assessment of surface cleanliness – Part 6: Extraction of soluble contaminants for analysis: The Bresle method. Geneva, Switzerland.

16. International Organization of Standardization (ISO) 8502 (2001). Preparation of steel substrates before application of paints and related products – Tests for the assessment of surface cleanliness – Part 8: Field method for the refractometric determination of moisture. Geneva, Switzerland.

17. International Organization of Standardization (ISO) 8502 (2020). Preparation of steel substrates before application of paints and related products – Tests for the assessment of surface cleanliness – Part 9: Field method for the conductometric determination of water-soluble salts. Geneva, Switzerland.

18. International Organization of Standardization (ISO) 8503 (2012). Preparation of steel substrates before application of paints and related products – Surface roughness characteristics of blast cleaned steel substrates – Part 1: Specifications and definitions for ISO surface profile comparators for the assessment of abrasive blast cleaned surfaces. Geneva, Switzerland.

19. International Organization of Standardization (ISO) 8504 (2019). Preparation of steel substrates before application of paints and related products – Surface preparation methods – Part 2: Abrasive blast cleaning. Geneva, Switzerland.
20. International Organization of Standardization (ISO) 8504 (2018). Preparation of steel substrates before application of paints and related products – Surface preparation methods – Part 3: Hand and power tool cleaning. Geneva, Switzerland.
21. International Organization of Standardization (ISO) 8504 (2021). Preparation of steel substrates before application of paints and related products – Surface preparation methods – Part 4: Acid pickling. Geneva, Switzerland.
22. International Organization of Standardization (ISO) 8504 (2019). Preparation of steel substrates before application of paints and related products – Surface preparation methods – Part 5: Water Jetting (Water Jet Cleaning). Geneva, Switzerland.
23. National physical laboratory (NPL). Guide to good practice in corrosion control. Surface preparation for coating.
24. Rust–Oleum (2021). Surface preparation. [online] available at: https://www.rustoleum.com/pages/industrialsolutions/resources/surface-preparation/surface-preparation-guide/ [access date: 11th April, 2021].
25. Van Dam, J.P.B., et al (2019). Effect of surface roughness and chemistry on the adhesion and durability of a steel epoxy adhesive interface. *International Journal of Adhesion and Adhesives*, Volume 96. pp. 2–5. Elsevier.

3 Coating Protection

3.1 COATING DEFINITION

Coating is a product in the form of a liquid, gas, solid or powder that is applied to the surface of an object, usually called the "substrate," to form a protective or decorative film. Coating may cover the whole surface or cover it partially. The primary objective of coating in the offshore industry is to protect metallic substrates from external corrosion. External corrosion is discussed in Chapter 1 and is caused by the corrosive offshore environment. The corrosion protection provided by coating involves three main features: The first is barrier protection, which reduces the ability of water, oxygen and contaminants from reaching the substrate, as illustrated in Figure 3.1. The second is inhibitive protection, in which chemicals such as lead and chromate are added to the coating to provide a layer between the coating and substrate and prevent their reaction, as illustrated in Figure 3.2. In addition to making a protective, inhibitive layer, red lead and chromate pigments are inhibiting pigments that can react to the binder of the coating and make a metal soap that acts as a basis for corrosion inhibition. The definitions of pigment and binder are given later in this chapter.

Lead and chromate are no longer used to enhance coating inhibition since these two elements have been found to be very toxic. The unfavorable health effects of these two compounds are not limited to toxicity. It has been shown that long exposure to coating that contains lead can cause motor neuron disease. Chromate can cause lung and throat cancer. Thus, modern pigments are lead- and chromate-free and can be either organic or inorganic. Inorganic pigments are phosphates, molybdates, silicates and ferrites. Organic pigments are based on the carbon chain and carbon rings. The third and final protection coating provides cathodic protection, in which the coating sacrifices itself to protect the steel. Figure 3.3 illustrates the sacrificial or cathodic concept of coating protection, which is associated with the presence of zinc.

While coating can be highly effective, damage to the coating can still occur. A "holiday" refers to a hole or void space or damage to the coating film that exposes the surface of the metal or substrate to corrosion attack. The definition of a holiday in coating is not limited to a hole or damage to the coating. A discontinuity or crack in the coating and improper adhesion or bonding of the coating are also considered a holiday. Figure 3.4 illustrates a holiday inside the coating. The sacrificial property of zinc offers a great advantage in protecting the substrate in the event of zinc coating damage. The other advantage of zinc in the case of holiday occurrence is to assist in the "healing" of the coating. The degree of sacrificial protection provided by the zinc depends on the thickness of the zinc coating and the corrosivity of the external environment.

DOI: 10.1201/9781003255918-3

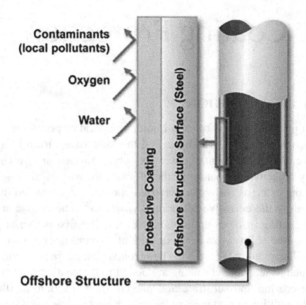

FIGURE 3.1 Coating barrier protection concept. (Photograph by author.)

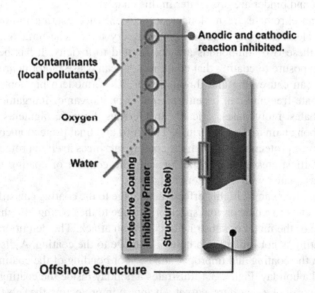

FIGURE 3.2 Coating barrier protection concept. (Photograph by author.)

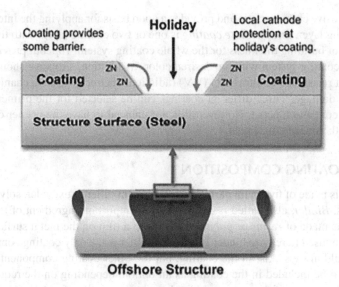

FIGURE 3.3   Coating barrier protection concept. (Photograph by author.)

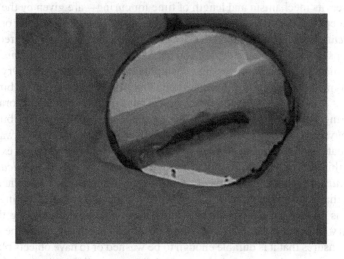

FIGURE 3.4   Holiday in the coating. (Photograph by author.)

What is the main difference between coating and varnish? Coating contains pigments that cause the formation of a blurred layer. The other essential product associated with surface protection is *varnish*, which provides a transparent film since varnish does not have any pigment. A coating typically consists of three main layers: *primer or undercoat, intermediate or tie coat* and *topcoat*. *Primer* is the first layer of coating on a substrate. It is expected to provide very good adhesion to the metal surface and increase the coating durability. Primer often has an anti-corrosive characteristic and provides a good basis for applying the intermediate coating layer. *Intermediate coating* is one or two coating layers with the main purpose of building thickness for the whole coating system. *Topcoat* provides the paint or coating system with the desired color. In addition, a topcoat should have sufficient resistance to ultraviolet (UV) radiation, chemicals and mechanical and abrasion damage. Three different coatings can be selected for the primer, intermediate coat and topcoat. However, it is not unusual to have the primer coat the same as the intermediate coat.

3.2 COATING COMPOSITION

Coating is made of five main parts: binder, pigment, filler or extender, solvent and additives. *Binder*, also called *resin*, is the most important ingredient of the coating and is made of *resins or polymers* that form a film on the metal surface. The other term used to refer to binder is *vehicle*. Binder is the only coating component that should always exist in the coating, but the other coating components listed above can be included in the coating as an option, depending on the required or expected property of the coating. Many key characteristics of a coating—such as adhesion to the metal surface; resistance to chemicals, the corrosive environment and water; its mechanism and length of time for curing—are given by the binder, which keeps the pigments together. Binders are dissolved in either water or chemical *solvents*, which can be called the carrier. Binders are divided into three different groups depending on the drying and hardening process, called *curing*. Curing is a chemical or physical process in which polymer materials harden. Drying and curing typically happen at the same time. The first type or group of binders is known as *oxidatively drying*; in this group of binders, the solvent evaporates and the coating absorbs oxygen from the air to dry. The second group of binders is called *physically drying* or *solvent evaporating*, in which the solvent or water part of the coating evaporates in order to dry the coat. Solvent for coating is explained later in this section. The third and last group of coating is called *chemically drying or curing* coating. In chemical curing or drying, there is a hardener in the coating that is mixed with the coating base to harden and cure the coating. When coating is applied on a metal surface in the form of a liquid, reducing the time necessary to dry and harden the coating poses a challenge. Curing the coating surface ensures that it is durable enough to be washed or to have objects placed on it, and that it is resistant to being scratched. There are different types of binders or polymer resins, such as *alkyd, acrylic, latex, phenolic, urethane* and *epoxy*. Phenolic and epoxy resins are two types of thermosetting polymers, meaning that they do not melt when they are heated.

Pigment is the next essential component of the coating. Pigments are solid, granular substances (see Figure 3.5) that are spread over the coating to give color to the coating and prevent transparency of the coating. In fact, pigment hides the substrate or metal surface. In addition, pigment protects the coating from UV light. The other effect of pigment is to change the coating properties; they increase hardness, decrease ductility, inhibit corrosion, add coating thickness and

FIGURE 3.5   Natural coating pigments. (Courtesy: Shutterstock.)

provide sagging resistance and gloss control. Sag resistance is the ability of coating or paint to resist sagging failure.

Sagging is a factor that is connected to coating composition and viscosity. *Sagging* refers to the downward drooping movement of the paint film immediately after application. One of the main causes of sagging is over-thinning of the paint. As a result of paint sagging, also called paint *running*, the paint moves downward after being applied to the substrate and before drying time, resulting in an uneven coating. Figure 3.6 illustrates variations in coating thickness caused by paint running or sagging.

The *gloss* or *sheen color* of a paint is related to the shining of the coating. Gloss color describes how much light is reflected from the surface of the paint. Smooth and glass-like paint is very shiny and glossy. The opposite of gloss is known as *flat* or *low-sheen* paint. One of the factors that affect coating gloss is pigment. Increasing the amount and concentration of pigments and extenders results in less gloss. In fact, the binder helps to create a smooth and glossy surface, and increase

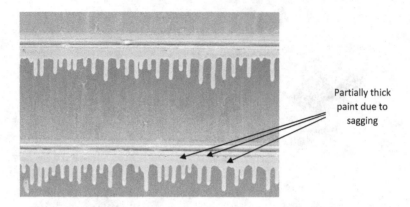

Partially thick
paint due to
sagging

FIGURE 3.6   Paint sagging. (Courtesy: Shutterstock.)

the amount and volume of pigment reducing binder volume. Figure 3.7 illustrates the distribution of pigments inside the resin or binder. The type of pigment used affects the gloss property of the paint. Pigments could be organic (natural) or inorganic (synthetic). Organic pigments are brighter and purer than inorganic pigments, but they are more prone to the risk of damage as a result of UV radiation. Coating gloss is measured by a gloss meter, as illustrated in Figure 3.8.

Titanium dioxide (TiO_2) is a widely used pigment type that is inorganic and white in color. Many types of pigment contain zinc. The main effect of zinc is galvanic protection, which is explained in more detail later in this chapter. Zinc, as a sacrificial metal, forms an anode and is corroded in favor of the metal surface in contact with it. Thus, the metal surface in contact with the zinc in the paint is protected. The effect of zinc on surface protection is called "cathodic protection" or "galvanic effect." Corrosion inhibition is generally associated with inorganic pigments, such as zinc oxide, zinc chromate and red lead. It is interesting to know that some types of coating could be pigment and/or solvent-free, but all types of coating contain binder or resin. When a coating is applied on a surface, the solvent evaporates during the curing process and only the resin (binder) and pigments are left on the substrate.

Solvent adjusts the curing properties and viscosity of the paint. Solvent is a volatile part of the paint that reduces the viscosity of the coating and does not become part of the coating film. Solvent affects the stability of the paint while the coat is in a liquid state. Solvent is the carrier of the non-volatile parts of the coating that dissolve the binder. *Thinner* oil, also simply called thinner, must sometimes be added to coating that is heavy or thick to make the heavy oil spread more easily on the surface. Thinner is a type of volatile solvent that is added to change a property, in this case the thickness, of the coat temporarily. The thinner evaporates after the coating has been applied on the surface and the paint has cured. As an oil-based solvent, thinner is used for diluting the coating. Alternatively, water can be used as a diluent in the coating, but in this case the water is not known as a solvent. Thus, two ways of making the coating thin are either by adding oil-based

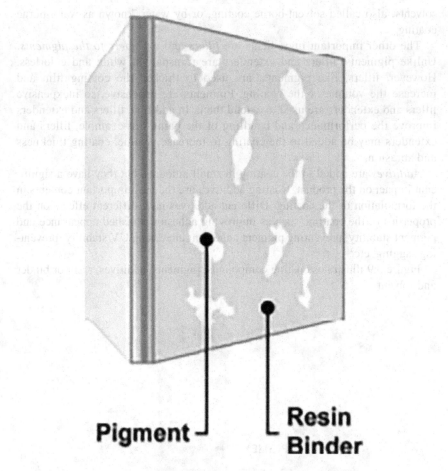

Pigment — **Resin Binder**

FIGURE 3.7 Distribution of pigments in the binder. (Photograph by author.)

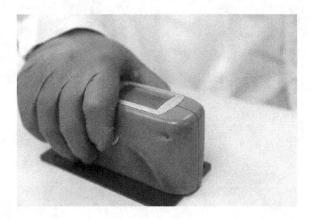

FIGURE 3.8 Coating gloss measurement by gloss meter. (Courtesy: Shutterstock.)

solvents, also called solvent-borne coating, or by water, known as water-borne coating.

The other important ingredients are *fillers* and *extenders to the pigments.* Unlike pigments, fillers and extenders are transparent, white and colorless. However, fillers, like pigments, are used to thicken the coating film and increase the volume of the coating. Pigments are expensive, so inexpensive fillers and extenders are used to extend them. In addition, fillers and extenders improve the performance and handling of the paint. For example, fillers and extenders may be added to the coating to increase volume, coating thickness and abrasion.

Additives are added to the coating in small amounts, but they have a significant impact on the product. Coating additives are the most important contents in the formulation of the coating. Different additives have different effects on the properties of the coating, such as improving adhesion, finished appearance and pigment stability; preventing pigment adhesion; increasing UV stability; preventing sagging, etc.

Figure 3.9 illustrates coating compounds, pigments, additives, resin or binder and solvent.

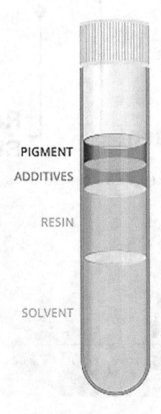

FIGURE 3.9   Coating components. (Courtesy: Shutterstock.)

3.3   COATING PURPOSES

The purpose of coating is not limited to corrosion protection. Coating, also called painting, can be used in the offshore oil and gas industry to provide identification; this is called *color coating*. Color coating applied on valves and actuators may convey a specific meaning from an engineering, fabrication or operation point of view. Color coding or marking will make identification easier for the application of the pipe and the fluid that is transported in it. As an example, valves and piping that are installed on a firewater system may be colored red. Figure 3.10 illustrates firewater piping and valves in red color. Red color piping systems, including industrial valves, have been used for firewater and firefighting installations across the globe for many years. The practice of identifying firewater piping systems with red color began in the USA. Other types of water piping systems inside plants, such as those used for drinking water or water injection, are not in red color. Firewater is a type of water used to extinguish fire, while drinking water is the source of water supply for the personnel who work in the plant; water injection piping contains water that will be injected into the oil reservoir to improve oil production when the reservoir pressure decreases after a period of time. Color coding these systems makes identification easy at a glance. Painting or coating can also be used for *decorative purposes*, which is more common in non-industrial applications.

The other important expectation from coating is its *anti-fouling* properties, especially for components located in the splash zone or submerged in

FIGURE 3.10   Firewater piping and valves in red color. (Courtesy: Shutterstock.)

FIGURE 3.11 Fouling of marine organisms on the bottom of a ship. (Courtesy: Shutterstock.)

seawater. The splash zone, explained in more detail in Chapter 1, refers to the area between the immersion and atmospheric zones that can get wet due to the splashing of seawater. All surfaces that are exposed to seawater are at risk of being attacked by marine organisms, an event called *biological fouling* or *biofouling*. In short, biofouling is defined as the settlement and growth of marine animals and plants on steel structures in or in contact with the sea or ocean, and fouling is defined as the undesirable deposition of materials on the surface. Fouling can be divided into organic or inorganic (chemical) fouling, particle fouling and biofouling. Figure 3.11 illustrates fouling by marine organisms on the hull of a ship.

One of the main negative impacts of biofouling is corrosion. The type of corrosion caused by biofouling can be called microbially induced corrosion (MIC). Biofouling can change the surface roughness of the metal, making it harder if it is not treated properly in the beginning. Poorly adherent biofouling products can be removed by light brushing or water spray. When biofouling attack occurs on the bottom of a ship, it can increase the surface roughness of the hull, resulting in slower movement of the ship and/or more fuel consumption. Fouling may occur internally in heat exchangers that use seawater as a medium for cooling warm media. One of the most effective ways to prevent biofouling attack on the external surface of a metal is to use proper coating on the metal surface. One of the coating types that is used for splash and submerged zones is two-component epoxy, which may be required to have anti-fouling properties.

3.4 COATING STANDARDS AND SPECIFICATIONS

The Society for Protective Coatings (SSPC) is an American National Standards Institute accredited standards development organization. The SSPC develops and publishes widely used industry standards for surface preparation, coating selection, coating application, painting contractor certification and testing. The SSPC designates different grades of surface preparation, which are explained in detail in Chapter 2. The National Association of Corrosion Engineers (NACE) international industry standards mainly address material selection and corrosion evaluation, but some of the NACE standards are related to industrial coating. The American Society for Testing and Materials (ASTM) is another international standard that addresses coating. ISO standards for coating are the other standards that are widely used, especially in Europe. Some of the main ISO coating standards are listed as follows:

ISO 8501: Visual assessment of surface cleanliness;
ISO 8502: Tests for the assessment of surface cleanliness;
ISO 8503: Surface roughness characteristics of blast-cleaned steel surfaces;
ISO 8504: Surface preparation methods;
ISO 12944: Paint and varnishes—corrosion protection of steel structures by protective paint systems.

The above-mentioned ISO standards are briefly reviewed in Chapter 2. NORSOK M-501, surface preparation and protection coating, is the main coating standard used in the Norwegian offshore industry. This standard was introduced in 1994. The NORSOK standards are developed by the Norwegian petroleum industry to ensure sufficient safety, value added and cost effectiveness for the oil and gas industry. There are different references to ISO standards in the NORSOK coating standard. In addition to the NORSOK coating standard, NORSOK M-501 identifies a series of coating requirements, including surface preparation, different coating material specifications and coating application procedures. A technical and professional requirement (TR 042), surface preparation and protective coating, has been developed by Equinor, a major operator company in Norway, and other project specifications may be used.

3.5   NORSOK COATING STANDARD M-501

3.5.1   GENERAL REQUIREMENTS

The selection of coating systems shall be made and documented in every project for required facilities and components such as valves and actuators. The manufacturer or supplier of items that require coating shall submit a coating procedure for approval. All revisions to the coating procedure should be documented during the project. Requirement and selection of a coating system and the applicable procedure shall consider different conditions, such as fabrication, operation and installation. It is important to keep a record and evaluate the team of personnel involved in the coating operation, such as management, operators and inspectors, as well as a list of coating facilities, before starting the job during the kick-off meeting.

The maximum environmental humidity for applying both blast cleaning and coating is 85%. Painting a metal surface at a high level of humidity above the provided limit can cause coating failure because of the high risk that the presence of moisture and humidity on the metal surface will cause rust eventually. Additionally, the steel temperature should be always at least 3°C above the dew point during both the sandblasting and coating operation to prevent humidity and condensation. The coating should be applied and cured only at ambient and steel temperature above 0°C.

The selection of coating materials should be undertaken by considering multiple parameters, such as corrosion protection properties, health, safety and environment (HSE) considerations, operating and application conditions (temperature, substrate material, etc.), economy and the track record of the coating. Coatings from new suppliers may require qualification. Additionally, some coating systems, such as systems 1, 5 and 7, must be prequalified by applying some qualification tests, such as an aging resistance test or seawater immersion test (applicable mainly to coating system 7). The prequalification requirement shall be fulfilled and documented before the start of any work. All the measurement and inspection records must be recorded and maintained.

A coating procedure specification (CPS) provides different information about surface preparation, coating, inspection, repair procedure, etc. A coating test procedure (CPT) is used to qualify the coating. Coating systems are explained in more detail later in this chapter. Topcoat colors are selected according to annex B of the standard and are listed in Table 3.1. RAL is the abbreviation of Reichs-Ausschuß für Lieferbedingungen und Gütesicherung. RAL is a European color-matching system that defines colors for paint.

Steel materials that require surface preparation should have at least rust grade B initial surface cleanliness as per both the NOROSK and ISO 8501-1 standards. This means that metal surfaces with rust grades A and B initial surface condition are acceptable for further surface preparation, and rust grades C and D should be avoided. Shop primer is defined as a protective coating normally applied for the protection of metal surfaces during transportation and storage. Zinc ethyl silicate primer in one coat with a thickness of 15 µm (and maximum 25 µm) could be applied as a shop primer. It is important to know that the proposed primer is applicable for steel surfaces with a cleanliness level of ISO 8501-1 Sa 2½.

Surface preparation in the NORSOK standard involves the removal of sharp edges, corners and welds, and smoothing by grinding to a minimum radius of

---

TABLE 3.1

Topcoat Color and RAL Designation as per
NOROSK M-501 Standard Appendix B

Color	RAL Designation
White	RAL 9002
Blue	RAL 5015
Gray	RAL 7038
Green	RAL 6002
Red	RAL 3000
Yellow	RAL 1004
Orange	RAL 2004
Black	RAL 9017

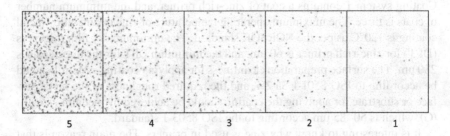

| 5 | 4 | 3 | 2 | 1 |

FIGURE 3.12 Dust particle size classes as per ISO 8502-3.

2 mm. In addition, hard surface layers (e.g., resulting from flame cutting) shall be removed by grinding prior to blast cleaning. Any grease or oil contamination shall be removed in accordance with SSPC SP1 before sandblasting by solvent cleaning, petroleum solvent, alkaline emulsion or steam cleaning. The surface profile should be in accordance with the ISO 8503 standard. Grit shall be used as a blast medium and shall be non-metallic and free from chloride. Two important final surface contaminants and conditions that should be inspected are dust and soluble salts and impurities. The accepted dust level is rating 2, as per ISO 8502-3. Figure 3.12 illustrates the different dust levels from 1 to 5, and the required sizes on the metal surface according to the ISO 8502-3 standard. Level 1 means that dust particles are not visible under X10 magnification. Level 2 indicates that dust particles are visible under X10 magnification but not with normal or corrected vision (usually particles <50 μm in diameter). Level 3 indicates that particles are visible with normal or corrected vision (usually particles between 50 and 100 μm in diameter). Level 4 addresses particle sizes between 0.5 and 2.5 mm in diameter. Level 5 addresses particles larger than 2.5 mm in diameter. The maximum content of soluble impurities on the surface shall not exceed 20 mg/m^2 according to the ISO 8502-9 standard.

3.5.2 Coating Systems

In summary, nine coating systems are defined in the NORSOK standard. However, the focus of this section is to introduce those coating systems that are applicable for piping, valves, actuators and accessories in the offshore industry. These are coating systems 1, 2 (2A and 2B), 5 (5A), 6 (6A, 6B, 6C) and 7 (7B, 7C).

3.5.2.1 Coating System 1: Organic and Inorganic Zinc-Rich Coating

Coating system 1 is applicable to the following materials and temperature range, as well as facilities and components:

- Carbon steel with maximum operating temperature ≤120°C;
- Steel structures, equipment, non-insulated piping and valves, actuators, gear boxes, etc.

Coating system 1 contains a coat of zinc-rich primer, and the minimum number of coats is three. The maximum operating temperature of zinc or zinc-alloy metal coating is 120°C as per the NORSOK standard. The minimum dry film thickness (DFT) for zinc-rich primer is 60 μm, and the minimum DFT for the three coats is 280 μm. The surface preparation cleanliness for applying coating system 1 should be according to ISO 8501-1 Sa 2½, and the required roughness of the metal surface or substrate for applying the coating should be achieved by medium-size grit (G), which is 50–85 μm according to the ISO 8503-1 standard.

It is interesting to know why zinc is used in primers. The main reason is that zinc acts as an anode or sacrificial metal, so it provides additional protection to the steel surface from corrosion called *galvanic protection*. The other main role of zinc is to increase the durability of the coating by providing corrosion protection against the corrosive environment through *barrier protection*. The main question is, how can zinc protect the metal surface? If you introduce a new, clean sample of zinc to the atmosphere, it will oxidize to form unstable water-soluble zinc oxide (ZnO) and zinc hydroxides ($Zn(OH)_2$). These two products are unstable and easily removed from the metal surface, so they cannot provide any barrier protection. The second compound, zinc hydroxide, is created as a chemical reaction between zinc and water. However, ZnO can make zinc carbonate ($Zn\ CO_3$), which is a stable layer and can provide barrier protection thanks to zinc. Barrier protection through the formation of zinc carbonate is known as the primary substrate protection, and the anodic characteristic of zinc provides the second barrier protection, galvanic protection.

Coating system 1 can be divided into two categories: coating systems 1A and 1B, according to TR042, the Equinor coating specification. The difference between 1A and 1B is that system 1A is NORSOK system 1 with an inorganic zinc-rich primer like zinc ethyl silicate, whereas system 1B is NORSOK system 1 with an organic zinc-rich epoxy primer. Adding epoxy to coating has some advantages, such as increasing the coating's resistance to chemicals and water, and improving coating adhesion. In addition, epoxy coating provides high mechanical strength. Epoxy coating has been used extensively for many years both onshore and offshore in the oil and gas industry to protect steel materials. One of the most important characteristics of epoxy resin is its ability to change a liquid coating to a resilient and solid one because epoxy is very viscous. However, one of the main disadvantages of epoxies is color change and chalking due to exposure to UV light. Chalking refers to the formation of powders on the paint surface. The binder or resin is degraded as a result of UV light, so some pigments loosen from the substrate and appear on the metal surface as dust. The other disadvantage of epoxy coatings is that they are in general thermosetting plastics or polymers, meaning that they get harder and become rigid as they are heated. There are two more disadvantages related specifically to zinc-rich epoxies; the first is that zinc epoxy coating cannot provide a thick film. Typically, zinc-rich epoxies can only provide a film thickness from 25 to 50 μm. In addition, zinc epoxy coating is not resistant to acids and can only be used in pH ranges of 6–9, typically. Zinc epoxy coating is widely used as a first coating or primer for topside or above-water coating. Zinc epoxy primer must be overcoated with other types of epoxy coating, as the active zinc is corroded easily, especially in a pH range of 6–9. The heat resistance of epoxy coatings is limited to 120°C. Aluminum silicate, which is an inorganic coating, can be used instead of epoxy coating for temperature ranges above 120°C. Figure 3.13 illustrates a small wedge gate valve coated with white (RAL 9002) zinc-rich epoxy coating. Valves are explained in more detail in Chapter 5.

Epoxy coatings can consist of one or two components. Coating system 1A, organic zinc-rich epoxy, is a one-component or one-part epoxy. The other type is called a *two-component epoxy*; its two parts, which are resin and hardener, should be mixed precisely in order to make the third chemical or final product. The final product or "third chemical" made by mixing resin and hardener is a type of thermosetting plastic. Two-component epoxy coatings can be cured at room temperature and do not necessarily need a hot temperature. The hardener itself triggers the required polymerization for curing. The chemical reaction between the hardener and resin in two-component epoxy leads to the formation of a coating product that is stronger than one-component epoxy. However, one of

FIGURE 3.13 Wedge gate valve coated with white zinc epoxy coating. (Photograph by author.)

the disadvantages of two-part epoxies is that they are dependent on the accurate mixing of the two components to achieve the required property. Two-component epoxies are used in coating systems such as NORSOK systems 5B, 6A, 6B and 7, which are explained later in this chapter.

The DFT of the coating specified in TR042 differs from the NORSOK standard. Tables 3.2 and 3.3 list acceptable suppliers, products and DFT ranges for systems 1A and 1B coating based on TR042, the Equinor coating specification.

All of the coatings in Table 3.2 are inorganic. *Carbozinc 11*, listed as the first coating product in Table 3.2, provides excellent corrosion protection for carbon steels in harsh offshore environments for over five decades. Carbozinc 11 is known as high-performance inorganic zinc protection for steel structures worldwide. The other advantage of this product is rapid curing, which means that it can be dried in 45 minutes in an environment with 50% humidity. Curing time is the amount of time it takes for a coating to react chemically with the metal surface, harden and eventually become dry. During the curing time, the coating should not be touched or exposed to water.

Resist 86 is another type of 1A or inorganic coating that can be cured very quickly in the presence of moisture. The material of resist 86 is inorganic zinc ethyl silicate. Inorganic zinc ethyl silicate can be used as a primer. In fact,

TABLE 3.2

Coating System 1A, Inorganic Zinc-Rich Coating, Accepted Suppliers, Products and DFT Range as per TR042, Equinor Coating Specification

	Coating System 1A		
Supplier	Product	Minimum DFT (μm)	Maximum DFT (μm)
Carboline	Carboline 11	75	150
	Carbo Guard E 19 primer	25	50
	Carbomastic 18 FC	130	300
	Carboxane 2000 TC	75	150
Hempel	Galvosil 15700	60	100
	Hampadur products	25	400
Intermational	Interzinc 2280	60	150
	Intergard 269	25	80
	Intergard 475HS	160	450
	Interfine 1080/2080	60	120
Jotun	Resist 86	75	90
	Pen guard Tie coat 100	25	50
	Jotamastic Plus	130	300
	Hardtop Optima	75	100
PPG	Sigmazinc 9	60	120
	Sigma fast 278 mist coat	25	40
	Sigma fast 278	140	500
	PSX 700	80	175

zinc-rich epoxy primer, which is required by NORSOK M-501, was widely used during the 1960s and 1970s. Later on, zinc ethyl silicate primers became more popular in the industry. Nowadays, new developments in zinc-rich epoxies have made this choice more popular again.

Carbozinc 858, listed as the first product in Table 3.3, is an organic zinc-rich primer and coating suitable for steel structures as well as actuators, piping and valves. This coating works properly in harsh and aggressive offshore environments and provides a hard, strong film.

What are the main differences between organic and inorganic zinc coatings?

1. *Binder:* The main difference between organic and inorganic coatings is the binder. An epoxy is often used for organic and a silicate for inorganic coating. Other types of binders, such as polyurethane and alkyd, could be selected for organic zinc-rich coatings. Epoxy coatings like zinc epoxy provide excellent chemical resistance with very good adhesion and water resistance.

2. *Durability:* The durability of inorganic zinc-rich coating is higher than organic zinc-rich coating because the galvanic protection effect is stronger in inorganic zinc-rich coating. As an example, organic zinc coating could have 13.5 years of durability, whereas the durability of inorganic zinc coating could be 15 years.

TABLE 3.3

Coating System 1B, Organic Zinc-Rich Epoxy, Accepted Suppliers, Products and DFT Range as per TR042, Equinor Coating Specification

Coating System 1B			
Supplier	Product	Minimum DFT (μm)	Maximum DFT (μm)
Carboline	Carboline 858	75	150
	Carbomastic 18 FC	130	300
	Carboxane 2000 TC	75	150
		75	150
Hempel	Hampadur Pro Zinc 17380	60	100
	Hampadur Mastic 4588W	160	400
	Hempaxene Light 55030	60	75
Intermational	Interzinc 52	75	200
	Interplus 356	150	400
	Interfine 1080/2080	60	120
Jotun	Barrier	75	90
	Jotamastic Plus	130	300
	Hardtop Optima	75	100
PPG	Sigmazinc 68SP	60	120
	Sigma fast 278	140	500
	PSX 700	80	160

3. *Skill requirement:* The application of an inorganic coating system requires more skill, care and attention.
4. *Topcoat application:* Topcoat application is easier with organic zinc systems. The topcoat is the last layer of coating applied on the metal surface. It is the first layer or line of defense against external objects and the corrosive environment. In fact, topcoat is the finishing sealer of the coat.
5. *Abrasion and sun resistance:* Inorganic sealants are highly resistant to abrasion, sun and solvents.

To summarize, inorganic zinc-rich epoxy is more common for applications of piping, valves and actuators in the offshore oil and gas industry according to the Equinor specification.

3.5.2.2 Coating System 2: Thermal Spray Aluminum or Thermal Spray Zinc

Coating system 2 is applicable for the following temperature ranges and materials:

- Carbon steel with maximum operating temperature >120°C;
- Insulated carbon steel in any temperature;
- Austenitic stainless steel with maximum operating temperature >60°C;
- Insulated duplex, super duplex and 6MO with maximum operation temperature >150°C;
- Uninsulated duplex stainless steel with maximum operating temperature >100°C;
- Uninsulated super duplex stainless steel with maximum operating temperature >110°C;
- Uninsulated 6MO with maximum operation temperature >120°C.

Coating system 2 is applicable to the following components and facilities:

- Tanks, piping and valves;
- Flares and crane booms: A flare boom is a structure that extends from the oil rig to provide safe burning of the gases in the flare system. The flare boom extends to provide a distance between the oil rig and the flare tip where the gases are burnt and released into the environment, as illustrated in Figure 3.14. An offshore crane is an elevating and rotating lifting device used to transfer materials and personnel to and from vessels and structures;
- The underside of the bottom deck, including piping; the jacket above the splash zone; lifeboat stations.

Coating systems are divided into two systems in the NORSOK standard. The first is system 2A, which is thermal-spray aluminum (TSA) or aluminum alloys. The minimum thickness of this coating is 200 μm. System 2A is the system used for piping and valves. The second is system 2B, which is thermal-spray zinc (TSZ)

FIGURE 3.14   Flare boom and flare system of an oil rig. (Courtesy: Shutterstock.)

or alloys of zinc with a minimum thickness or DFT of 100 μm. Coating system 2B, metallic zinc, is not common, unlike the 2A coating system with metallic aluminum. The reason is that coating system 2 in general is widely used for different types of stainless steel, and coating for stainless steel must not contain metallic zinc. In fact, the melted zinc used in coating system 2B can corrode stainless steel. Corrosion tests and data show that molten zinc at a temperature of 500°C or higher is very aggressive to both ferritic and austenitic stainless steels in even short-term contact. As an example, stainless steel type 316 can be corroded significantly because of contact with molten zinc; this is called liquid molten metal zinc corrosion. The resistance of 6MO as a type of super austenitic is much higher than austenitic stainless steel 316 to the molten zinc. Coating system 2A suppliers, products and minimum thickness as per the Equinor coating specification are provided in Table 3.4. The required substrate surface cleanliness and roughness before applying coating system 2A are the same as for coating system 1; this means that the cleanliness is required to be ISO 8501-1 Sa 2½, and the roughness should be achieved by medium-size grit (G), which is 50–85 μm according to ISO 8503-1.

System 2, which includes both TSA and TSZ, also called *metallization* or *metal coating*, is a process in which powder or a metal wire is melted and then sprayed on a metal surface for corrosion protection. Flame spray, which is a mixture of oxygen and a choice of fuel gas such as propane, is used to generate flame at the tip or nozzle of the gun-shaped torch. Wires fed into the rear part of the nozzle are melted by the source of energy. The melted particles or atoms are accelerated toward the substrate surface by the gas flow and air jet. The air jet produced by the compressor increases the velocity of the thermal spray. Figure 3.15 illustrates zinc metal coating on a metal surface. This type of coating is not proposed for stainless steel, because melted zinc causes liquid molten metal zinc corrosion of stainless steel, as explained above.

TABLE 3.4

Coating System 2A Accepted Suppliers, TSA Products and DFT Range as per TR042, Equinor Coating Specification

Coating System 2A		
Thermal Spray Aluminum (TSA)		
Supplier	Product	Minim Thickness
Carboline	Thermaline 4674	200 μm with sealer
Hempel	Silicone Aluminum 56914	200 μm with sealer
International	Intertherm 50	200 μm with sealer
Jotun	Solvalitt	200 μm with sealer
PPG	Sigmatherm 540	200 μm with sealer

Gun shape torch

Melted zinc metal spray and flame

FIGURE 3.15   Metal coating with zinc. (Courtesy: Shutterstock.)

Coating system 2, which includes both TSA and ZSA, is an extremely durable coating system that provides significant protection and lifecycle improvement compared to other types of coating systems, such as coating system 1. It has been reported that metal coating, meaning coating system 2, can be used in tidal and splash zones with a durability of 30 years. This type of coating is so strong that it does not require the addition of any cathodic protection.

Coating system 2 is a good choice for a variety of applications, such as marine, subsea, high-temperature, etc. The advantages of this type of coating can be summarized as follows: a minimum need for maintenance; low cost, considering both initial and operation costs; superior adhesion to the metal surface; resistance to mechanical damage and to solvents; no required curing time, meaning that the component coated with TSA can be handled soon after finishing the coating. In addition, it covers a wide range of operational temperatures, from −45°C to 538°C, and the presence of aluminum in the coating system as an active and sacrificial metal provides good protection for steel. All metalized surfaces are porous,

FIGURE 3.16 A modular valve coated with TSA for offshore use. (Photograph by author.)

so they should be filled in with a proper sealer as per the NORSOK M-501 standard. This standard proposes two-component epoxies for operating temperatures <120°C, and aluminum silicone for operating temperatures above 120°C as sealer materials for coating system 2A. Figure 3.16 illustrates a modular valve coated with TSA. More details about the valves used in the offshore oil and gas industry are provided in Chapter 5.

3.5.2.3 Coating System 5: Passive Fire Protection Epoxy or Cement

Fire is the most common major accident in process plants, as illustrated in Figure 3.17. Fire requires three elements: fuel, heat and oxygen. Fire is very hazardous in industrial plants, as it can produce live flames, sparks and hot objects that may interact with chemicals that have the potential for ignition or that can intensify the fire such that it becomes large and uncontrolled. There are five classes of fire based on types of fuel:

- *Class A*: Fuel includes ordinary combustibles like wood, paper, plastic or anything that leaves ash;
- *Class B*: Fuels include petroleum, oil, paint or flammable gasses;
- *Class C*: Fuel is the ignition of electrical components like motors;
- *Class D*: Fuel includes combustible materials like potassium, sodium, aluminum, etc.;
- *Class K*: Fuels include cooking oils and greases such as animal and vegetable fats.

Fires inside offshore plants most often belong to class A and B. Fire class A, with wood and paper, can reach 556°C in just 5 minutes and to 945°C after 1 hour.

FIGURE 3.17　Offshore platform fire. (Courtesy: Shutterstock.)

Fire class B, with oil and hydrocarbon, can reach 926°C after 5 minutes and 1,145°C after 1 hour. It is important to know that steel will lose its strength in a temperature range of 550°C–625°C. Fire not only produces a high amount of heat but can cause a high amount of load and lead to explosions that are crucial to consider from a safety engineering point of view. Fire load is defined as the weight of combustible material per applied area. Additionally, ignited release of pressurized and flammable fluid can cause a jet fire with significant momentum in a specific direction. Thus, oil and gas plants, including offshore assets, are at risk of explosion and fire. These negative incidents can escalate rapidly, resulting in catastrophic events, such as loss of life and assets, if passive fire protection measures are not in place. An organic product (epoxy intumescent coating) or inorganic product (cement) can be used for coating to protect against fire; these coatings are explained in more detail below.

NORSOK coating system 5 is divided into two categories: coating system 5A, epoxy-based fire protection; and coating system 5B, cement fire-based protection. NORSOK system 5A is more common than system 5B in the offshore oil and gas industry. Epoxy-based fire protection, also called intumescent epoxy, is a type of fire-resistant epoxy coating system that can be applied directly on the substrate to provide both fire and corrosion protection and act as passive fire protection (PFP). In fact, fire-resistant epoxy materials expand when in contact with fire to form an insulated char. This char has low thermal conductivity, so it acts as a thermal barrier between the fire and the substrate. According to Det Norske Veritas (DNV) standard DNV-OS-D301, fire standard for offshore, PFP could be a coating or a free-standing system that provides thermal protection in case of fire. DNV is an international company with headquarters in the Oslo area that provides various services in the oil and gas industry and other sorts of industries like renewable energy. PFP as a system or unit is a type of enclosure in the form of a metallic box (rigid enclosure) or flexible jacket. PFP can be installed on piping, valves,

FIGURE 3.18   Passive-fire protection (PFP) around a valve actuator and control panel. (Picture by author.)

actuators, equipment and supports to protect this equipment from fire. Figure 3.18 illustrates PFP installed around a valve actuator and control panel. The actuator is hydraulic, and the control panel has electrical components that are ignitable, so PFP is applied around them to protect them from fire. An actuator, explained in more detail in Chapter 5, is an electrical or mechanical device responsible for opening and closing a valve. A control panel is a box that transports and commands the hydraulic fluid to the actuator. Different devices, such as valves, tubes, filters and pressure gauges, are installed on control panels.

One of the most important features of fire-resistant epoxy coating is that it requires mesh. Mesh provides strength for the coating by preventing char fall, detachment and cracking. It also provides protection against the erosive forces of jet fire. The protective effect of mesh is different in cement-based and epoxy-based fire-resistant coatings. Cement-based coatings do not expand in the event of fire, and they are thick, so they need some kind of support to keep them attached to the substrate. Wire mesh is a very common type of PFP coating reinforcement. It is necessary to affix the mesh to the steel substrate with pins or screws that are properly embedded into the PFP coating materials, which requires additional labor work. According to the NORSOK standard, the material of wire mesh should be hot-dip galvanized (HDG; zinc galvanized) or stainless steel materials for epoxy-based PFP coating, and plastic coated for cement PFP coating. Figure 3.19 illustrates a zinc-coated wire mesh. Using wire mesh is proposed by the NORSOK standard and is a traditional coating reinforcement. There are challenges and risks associated with wire mesh application, such as complex and time-consuming installation, the availability of personnel to do the task and the event of possible errors in installation. Many statistics indicate that errors in the installation of the meshes could lead to loss of life, loss of asset, etc. Thus,

FIGURE 3.19   Zinc-coated wire mesh. (Courtesy: Shutterstock.)

mesh-free, epoxy-based PFPs have been developed and used in the industry to prevent the disadvantages related to wire-mesh installation.

The surface preparation for coating system 5 is the same as for coating systems 1 and 2; the cleanliness level should be Sa 2½ according to ISO 8501-1. The roughness of the surface should be as per ISO 8503 standard which is achieved by medium-size abrasive grit (grade G) equal to 50–85 μm. One coat epoxy primer in 50 μm or one coat zinc-rich epoxy in 60 μm and a one-layer epoxy tie coat should be used in 25 μm thickness. Two layers of epoxy coating should be followed by epoxy fire-based protection. It is important to note that the dried thickness of the coating should be at least 85 μm. Cement-based fire protection coating (system 5B) is thicker than system 5A and should have a minimum thickness of 260 μm.

3.5.2.4   Coating System 6: Coating on HDG or Phenolic Epoxy

Coating system 6 in the NORSOK standard is categorized into three types of coating subsystems: Coating system 6A is used for uninsulated stainless steel or aluminum when painting is required. As an example, stainless steel 316 piping and valves may require coating system 6A for operating temperatures above 60°C. As mentioned above, NORSOK system 2A is an alternative coating system for stainless steel 316 above 60°C. Practically speaking, this type of coating is not common for valves and actuators in the Norwegian offshore industry, as stainless steel 316 is not typically used for operating temperatures higher than 60°C. The surface preparation for applying coating system 6A is different from the other coating systems discussed here and requires sweep blasting with non-metallic, chloride-free grit to achieve a surface profile of 25–80 μm. The coating consists of three layers: one coat of epoxy primer with a minimum DFT of 50 μm for the first layer. The second or intermediate layer is one coat of two-component epoxy with a minimum DFT of

100 μm that is finished with a minimum 75 μm topcoat. The minimum thickness or DFT of the complete coating system should be 225 μm.

Coating system 6B is not applicable for actuators but could be used on some parts of valves and piping such as bolting (bolts and nuts). This type of coating is applied on components that are coated with HDG or zinc coating. HDG is a form of galvanization, a process of coating iron and steel with zinc. The process of galvanization is performed by immersing the metal in a bath of molten zinc. Figure 3.20 illustrates the process of galvanizing steel components in molten zinc.

The only components in piping and valves that are galvanized are bolts and nuts. Typically, bolts and nuts for valves and flanges remain uncoated if they are galvanized. As an alternative, if the bolts and nuts are coated, the same coating that is used for the whole pipe or valve is used for the bolting. Figure 3.21 illustrates uncoated, galvanized bolts and nuts on a coated flange in a piping system. Figure 3.22 illustrates a carbon steel body valve coated with NORSOK system

FIGURE 3.20 Steel components in a molten zinc bath (galvanizing process). (Courtesy: Shutterstock.)

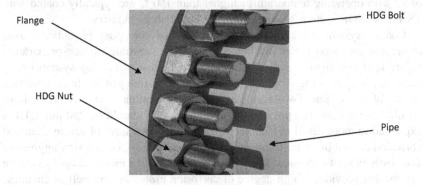

FIGURE 3.21 Galvanized bolts and nuts on a coated flange on a piping system. (Courtesy: Shutterstock.)

Coated
bolting

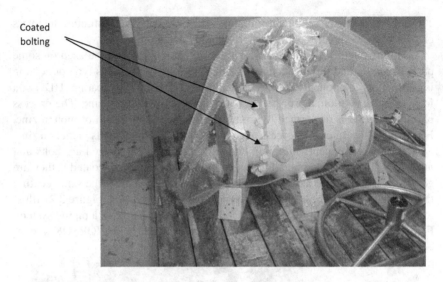

FIGURE 3.22   A carbon steel body valve and bolting coated with zinc epoxy. (Photograph by author.)

1A, zinc-rich epoxy, for offshore use; the bolting is coated with the same zinc-rich epoxy.

Surface preparation before applying coating system 6B requires cleaning with alkaline detergent followed by hosing with fresh water. The coating layers and thickness for coating system 6B is same as for system 6A.

The most important subsection of coating system 6 is coating system 6C, which is applicable for stainless steel materials that are insulated and have operating temperatures <150°C. As examples, insulated piping and valves in austenitic stainless steel, duplex, super duplex and 6MO with an operating temperature <150°C, as well as insulated actuators, accumulators and gearboxes can be coated with coating system 6C. As mentioned above, insulated duplex, super duplex and 6MO with operating temperatures higher than 150°C are typically coated with TSA, NORSOK system 2A, in the Norwegian offshore industry.

Coating system 6C, known as a phenolic epoxy or epoxy phenolic, is used to prevent corrosion under insulation (CUI). The substrate surface preparation before applying coating system 6C is similar to that of coating system 6A; it requires sweep blasting with non-metallic, chloride-free grit to obtain a surface profile of 25–85 μm. Two-layer phenolic epoxy coating is applied; each layer should have at least 125 μm and the maximum DFT should be 250 μm. CUI is explained in more detail in Chapter 1. It is defined as a form of severe localized corrosion caused by water trapped under the insulation. CUI is a very aggressive and costly type of corrosion. Phenolic epoxy coating is a modified epoxy coating system that provides a high degree of corrosion protection, as well as chemical and water resistance for wide ranges of applications. Two main advantages of phenolic epoxy coating are that it can be used in low pH (acidic) environments and high-temperature applications where high heat resistance is required. Other properties of epoxy that can be improved by adding phenolic are its excellent solvent resistance and hard, abrasion-resistant film.

3.5.2.5 Coating System 7: Two-Component Epoxy

Coating system 7 is divided into three categories: 7A, 7B and 7C. System 7A is used for carbon steel and stainless-steel coating in the splash zone. The splash zone is an area in the sea or ocean located immediately under and above the mean seawater level. This area is at major risk of corrosion in the offshore environment, as discussed in more detail in Chapter 1. System 7A is not typically applicable for valves and actuators, since they are not installed in the splash zone. The cleanliness level of the substrate before applying coating should be ISO 8501-1 Sa 2½. The roughness of the surface should be based on ISO 8503 standard achieved by medium-size abrasive grit (G), equal to 50–85 μm. Coating 7A could be either two-component epoxy or polyester-based coating with a minimum number of two coating layers. The complete coating system should have a minimum 600 μm DFT. Polyester-based epoxy and silicate have a chemical curing mechanism. The main reason why polyester coating is proposed for the splash zone is that splash areas are exposed to a high degree of erosion and wearing. Polyester coating is a thick coating type with excellent wearing and abrasion resistance. In addition, the curing time of polyester is fast (a few hours), and it has very good corrosion and chemical resistance.

NORSOK system 7B is applied on submerged carbon and stainless steel materials with operating temperatures of a maximum of 50°C. The cleanliness level of the substrate before applying coating should be ISO 8501-1 Sa 2½. The roughness of the surface should be based on ISO 8503 standard achieved by medium-size abrasive grit (G), equal to 50–85 μm. The surface preparation, including the cleanliness and roughness of the substrate before applying coating system 7B, is the same as for coating system 7A. Coating system 7B includes two-component epoxy with a minimum number of two layers and a minimum thickness of 350 μm DFT. Therefore, the film thickness for coating system 7B is less than coating system 7A.

Coating system 7C is used for submerged carbon and stainless-steel materials in operating temperatures higher than 50°C. The surface preparation, including the cleanliness and roughness of the substrate before applying coating system 7C, is the same as for coating system 7A and 7B. The cleanliness level of the substrate before applying coating should be ISO 8501-1 Sa 2½. The roughness of the surface should be based on ISO 8503 achieved by medium-size abrasive grit (G), equal to 50–85 μm. Coating system 7C, like 7B, requires at least two layers of coating in two-component epoxy with a minimum DFT of 350 μm. Two-part or two-component epoxy was explained earlier in this chapter.

There are some important points to note with regard to coating system 7; the first, which is in general correct for the sandblasting of stainless steel, is that the abrasives should be chloride-free and non-metallic. The second point is that a coating system in a submerged zone is always combined with cathodic protection.

Since most types of coating are imperfect, cathodic protection as a secondary external surface protection can be used in conjunction with coating in subsea environment. The third point is that an anti-fouling coating type may be required for submerged zones to prevent organic growth or the formation of deposits or microorganisms such as bacteria.

3.6 COATING HEALTH, SAFETY AND ENVIRONMENT (SHE)

There are potential hazards associated with painting implementation that jeopardize safety, health and the environment (SHE). In general, there are fewer health hazards associated with water-based compared to oil-based coating. Some of the potential hazards are listed as follows:

1. Rashes and swelling on the skin due to contact with the paint;
2. Dizziness from a short period of inhaling, eye irritation, sore throat, cough, fatigue and vomiting;
3. Solvents that are evaporated from the coating can cause headaches and drowsiness. In addition, solvents can be absorbed by the skin and cause skin degreasing and eczema. Solvents have negative impacts on the liver, kidneys, respiratory organs, blood and nervous system. Solvents contain different types of volatile organic compounds (VOCs). VOCs are compounds with high vapor pressure and low water solubility, and many of them are human-made. The emission of VOCs, also called fugitive emission, into the environment is undesirable as it causes environmental pollution. The other important consideration about coating solvent is that it is heavier than air and can spread on the floor and catch fire eventually. Solvents can enter the body via inhalation, skin contact and ingestion;
4. Long-term exposure to coating products can lead to liver, kidney, lung, digestive system and central nervous system damage, as well as cancer.
5. Coating can cause fire, especially if it is applied in unventilated areas. Coating areas should be free from ignition sources, such as cigarettes, sparks, static electricity and high heat. Solvent-based coating requires particular care with regard to safety issues, as the solvent that evaporates from the coating is flammable. The fire can cause blasts and explosions consequently. The flammability of the paint depends on the *flash point*. Flash point is defined as the lowest temperature at which a chemical can vaporize to form an ignitable mixture in the air. As a general rule, lower flash point means higher flammability. There are three classifications of flash point and flammability: extremely flammable with a flash point <0°C; highly flammable with a flash point from 0°C to 21°C; flammable with a flash point between 21°C and 55°C. The flash point and flammability of a given coating can be found on the material safety data sheets (MSDS). Coatings have MSDSs that contain information about their chemical composition, health and environmental hazards and protective measures. As an example, Figure 3.23 shows a caution sign with regard

FIGURE 3.23 Label warning of low flash point and flammability for a paint. (Courtesy: Shutterstock.)

to the low flash point and flammability of coating that can be found on the paint tin and the MSDS. Some of the information that is typically included in the coating MSDS is the product name and the company that produces the coating; hazards associated with the product, such as toxicity, flammability, corrosivity, etc.; coating composition and ingredients; exposure control and protection; first aid measures with regard to contact with the paint; safety measures with regard to coating fire; handling and storage; physical and chemical properties; transportation information, etc.

6. Epoxy binder coatings can cause eye and skin irritation and skin allergies. In fact, epoxy coatings can release VOCs during their application; these VOCs can contribute to smog and acid rain and cause pollution and damage to peoples' lives and to the environment. The odor of epoxy is due to its solvent. VOCs are part of the epoxy coating solvent, so if the epoxy coating does not have any solvent and is water-borne, it can be almost VOC free. Solvent-free epoxy, called solid epoxy resin, can be used to reduce the health hazards associated with other forms of epoxy.

HSE concerns are also related to coating with heavy metals. Some of the metals used for coating include chromium, iron, lead, cadmium, cobalt, copper, zinc and nickel. Contact with metals could cause harm to the skin, respiratory organs, kidneys, heart, etc. Dust, which is a surface contaminant, is another source of concern for workers' health, especially during and after sandblasting. Inhaling dust can create long-term damage. The other health problem posed by surface preparation or steelwork is noise. The noise levels during steelwork or surface preparation could go beyond the level that can be managed by using ear protectors. In addition, blasting and water jets produce high-pressure applications of sand and water that should always be kept away from personnel.

The location of the paint job has a direct impact on safety considerations. There is a big difference between working outside and working in a closed space, such as a coating room or coating booth. When coating is applied outside in the open air, the fresh and open air is usually sufficient to prevent coating vapor accumulation at a harmful and unhealthy level. However, attention must be paid when applying coating inside a coating booth. It is essential to make sure that the ventilation is working properly before beginning the coating process. Nothing that can create sparks should be used inside a coating booth. Large amounts of coating and unneeded coating should not be placed in the booth. Portable lamps and heaters should be kept out of the closed painting area. The enclosed coating area should have fire extinguisher and/or sprinkler. Any personnel working in the booth should wear a respirator, which is like a mask, with sufficient air supply over the mouth and nose to prevent the inhalation of unhealthy substances. The coating MSDS must be read carefully by the personnel applying the coating. The clothing of the worker should fully cover the skin and should be antistatic and include a hood. Safety boots and gloves are mandatory. Figure 3.24 illustrates a painter coating a pipe on an offshore platform with full safety protection.

Masks that provide respiratory protection against dust come in three grades: filtering face piece (FFP) types 1, 2 and 3. These three levels of mask protection against dust are shortened to P1, P2 and P3. FFP1 or P1 dust masks provide the lowest level of protection against dust and provide protection against suspended solids and liquids in the air. FFP1 masks can be used for operations such as cutting and sandblasting with sandpaper. FFP1 can filter at least 80% of airborne

FIGURE 3.24   A worker coating an offshore pipe. (Courtesy: Shutterstock.)

particles. FFP2 provides a moderate level of protection against the dust created by sandblasting. FFP2 can filter at least 94% of airborne particles. FFP3 is the highest level of dust protection and can be used for handling hazardous powders and solids. FFP3 can filter at least 99% of airborne particles. In addition, there are three levels of respiratory protection against the gas from organic solvents: A1 (the lowest degree of protection), A2 (medium degree of protection) and A3 (highest degree of protection).

Three essential actions should be taken if a worker is affected by paint hazards due to inhalation, or skin or eye contact. In the event of inhalation, the person should go to fresh air. In some cases, oxygen or artificial respiration may be required. Any skin affected by paint should be washed with soap and water. Any eye that comes in contact with paint should be washed with warm water for at least 15 minutes, and extra medical help is required as well.

3.7   CONCLUSION

This chapter explains different mechanisms of metal surface protection provided by coating. Coating compositions and formulations are discussed in more detail. Different coatings like zinc epoxy, inorganic zinc, TSA, phenolic epoxy, two-component epoxies, etc. that are used for piping and valves in the offshore industry are covered in detail in this chapter.

3.8   QUESTIONS AND ANSWERS

1. Which sentence is correct about coating and surface preparation?
    A.   Surface preparation is done prior to the application of coating, just to clean the surface.
    B.   Coating is used as a secondary means of protecting metal surfaces from external corrosion in combination with cathodic protection.
    C.   All coatings are eventually damaged because of contact with oxygen, water and chemicals, and by abrasion, pressure and temperature fluctuations.
    D.   Coating selection is more important than surface preparation for coating durability, adhesion and performance.
    Answer: Option A is not correct, because the main purpose of surface preparation is not limited to surface cleaning. In addition to cleaning, preparing the surface by reducing its roughness, removing sharp edges, grinding hard surfaces and in general proper preparation of the surface profile should be implemented during surface preparation. Option B is not correct, since coating is known as the primary surface protection and cathodic protection is used as the secondary metal surface protection. Option C is correct; all types of coating are subject to damage and failure eventually. Option D is not correct, because surface preparation is the most important factor in coating durability, adhesion and performance.

2. Which of the following sentences are correct regarding coating composition?
   A. Resin or binder is dissolved in water or a chemical solvent; the solvent is non-volatile, and the binder is the volatile part of the coating, which evaporates during the curing process.
   B. The adhesion of the coating to the substrate and coating corrosion resistance against the external environment or weathering are related to the binders.
   C. Coating thickness, the protection of the coating against UV light, sagging and gloss color are dependent on the pigments.
   D. Fillers and extenders to the pigment are colors added for economic reasons; they add thickness and volume.

   Answer: Option A is not completely correct. The resin or binder is dissolved in water or a chemical solvent, but the water or chemical solvent is the volatile part that is evaporated from the substrate during the curing process. Therefore, the binder is non-volatile, and the solvent is volatile. Options B and C are correct, but option D is not completely correct. Fillers and extenders to pigments are white, i.e., not colorful, but fillers and extenders are added to coating to provide some properties that are typically given by the pigment, such as thickness and volume. The reason why fillers and extenders are added is that pigments are expensive. Therefore, only two options, B and C, are completely correct.

3. Which sentences are not correct regarding curing the coating?
   A. Curing and drying are completely the same.
   B. Curing and drying do not happen at the same time.
   C. During physical curing, the coating gets dried and hardened by the evaporation of the solvent.
   D. Curing takes place before applying the coating to the substrate.

   Answer: Option A is not correct, because curing and drying are not completely the same. Curing is a physical or chemical process in which the coating gets both hard and dry. Thus, curing the coating involves more than just drying it. Option B is not correct, because curing and drying happen at the same time. Option C is correct; physically curing means that the coating gets dried and hardened (cured) by the evaporation of the solvent or water. Option D is not correct, as curing takes place after applying the coating to the substrate. Thus, all the options are wrong except for option C.

4. Identify the incorrect sentences regarding coating system 1 in the NORSOK M-501 standard.
   A. Zinc cannot provide any protective barrier in coating, since the layer of ZnO is unstable.
   B. Zinc epoxy is an organic coating that is made by mixing at least two components.

C. The surface preparation cleanliness grade of the substrate for applying zinc epoxy or zinc ethyl silicate primer should be at least ISO 8501-1 Sa 2 ½ as per the NORSOK standard.

D. Zinc epoxy is organic and does not pose any problem for the environment or peoples' lives.

Answer: Option A is not correct. Zinc can make a stable layer of zinc carbonate on a metal surface, which provides a protective barrier. Option B is not completely correct, as zinc epoxy is an organic coating made by either one component or by mixing exactly two components. Option C is completely correct. Option D is also wrong, as zinc epoxy implementation on a metal surface can release many undesirable compounds to the environment that pollute the air and put human lives in danger.

5. An uninsulated valve in duplex material has an operating temperature of 125°C. Which coating best suits this application?
   A. Zinc-rich epoxy
   B. Thermal-spray aluminum (TSA)
   C. Thermal-spray zinc (TSZ)
   D. Phenolic epoxy

   Answer: Option A is not correct for two main reasons: first, zinc-rich epoxy, which is NORSOK system 1, is applicable for carbon steel materials and not stainless steel because of the possible corrosion of stainless steel in contact with zinc. Second, zinc epoxy coating is suitable for a maximum operating temperature of 120°C, while the operating temperature in this application is 125°C. Option B, TSA, is the correct answer. TSZ is not proposed for coating stainless steel material, as the molten metallic zinc that is sprayed on the substrate causes liquid zinc molten metal corrosion of stainless steel, so option C is not correct. Option D, phenolic epoxy is not correct, as this type of coating is mainly used for insulated stainless steel materials. Thus, option B is the correct answer.

6. Which words should be placed in the blanks?

   _____ is called metalizing, in which powder or a metal wire is melted and sprayed on the metal surface for corrosion protection.

   _____ is a modified epoxy coating mainly used to prevent corrosion under insulation; it provides a high degree of corrosion protection, as well as chemical and water resistance. The other coating system used for passive fire protection is _____.
   A. Thermal-spray aluminum, zinc-rich epoxy, cement
   B. Thermal-spray zinc, phenolic epoxy, fire-based epoxy
   C. Thermal spray (zinc or aluminum), silicate epoxy, cement-based or epoxy-based fire coating
   D. Thermal-spray zinc or aluminum, phenolic epoxy, cement-based or epoxy-based fire coating

Answer: Option A is not completely correct. TSA is a type of metalizing, but zinc-rich epoxy is not a type of modified epoxy coating used to prevent CUI. Alternatively, phenolic epoxy is widely used to prevent CUI. Cement, the last word in option A, is not a coating system. Option B is correct, but it is not complete because both TSA and thermal spray zinc are metalizing. Option C is not completely correct, because silicate epoxy is not suitable for CUI. Option D is the correct and complete answer. TSA or TSZ are both called metalizing, as molten zinc or aluminum may be sprayed on the substrate. The other important point is that both cement-based and epoxy-based fire coating are used for PFP.

7. Which sentences are not correct regarding coating system number 5?
   A. Coating system number 5 indicates only an epoxy fire-resistant coating as a type of PFP.
   B. One of the advantages of PFP coating is the use of mesh to prevent them from falling and becoming cracked.
   C. PFP can be a component in the shape of a rigid box or flexible insulation rather than a coating as per DNV-OS-D301, the fire standard for offshore.
   D. The substrate surface preparation before applying coating system number 5 should be ISO 8501-1 Sa 2½. The roughness of the surface should be based on ISO 8503 achieved by medium-size abrasive grit (G), equal to 50–85 µm as per NORSOK standard.

   Answer: Option A is not correct, as coating system 5 addresses both epoxy fire-resistant coating and cement-based fire-resistant coating. Option B is not correct; using mesh on coating system 5 to prevent it from falling or becoming cracked is not considered an advantage for two main reasons; the first is that an error in the installation of the mesh could lead to coating failure and very negative consequences like fire, explosion, loss of life and environmental pollution. The second disadvantage of mesh installation is the additional labor it requires and the consequent increase in cost. Option C is correct (refer to DNV-OS-D301, the fire standard for offshore); PFP could be in the form of a coating or a free-standing system. Free-standing PFP could be a firebox or insulation coating. Option D is correct; refer to NORSOK M-501, coating standard. Thus, options A and B are not correct.

8. Which of the following sentences are correct with regard to NORSOK system 6?
   A. Aluminum or uninsulated stainless steels may be coated with NORSOK system 6A, but this type of coating is not common for piping, valves and actuators.
   B. Coating system 6B is used over galvanized stainless steel.
   C. Coating system 6C, phenolic epoxy, is applied to insulated duplex stainless-steel valves with an operating temperature of 160°C.

D. Coating system 6C, phenolic epoxy, has higher corrosion and chemical resistance compared to coating system 1A (zinc-rich epoxy).

Answer: Option A is completely correct. Option B is not correct;
galvanizing stainless steel is a wrong approach, as molten zinc coating or galvanizing can corrode stainless steel. In fact, coating system
6B is applied on galvanized steel, and not galvanized *stainless* steel.
Option C is not correct; the maximum operating temperature for phenolic epoxy is 150°C, so it cannot be used for an operating temperature
of 160°C. TSA, coating system 2A, is recommended for an insulated
duplex stainless steel valve with an operating temperature higher than
150°C, as in this case. Option D is correct, since phenolic epoxy has
higher corrosion and chemical resistance compared to coating system
1A. In fact, CUI is more critical than offshore atmospheric corrosion.
Therefore, phenolic epoxy coating, which is applicable for CUI, should
be stronger than zinc-rich epoxy, which is used for protection against
the corrosive offshore environment. In conclusion, options A and D are
correct.

9. A super duplex valve is installed at a depth of 2,300 m in the sea. The
operating temperature of the valve is 70°C, and the media inside the
valve is a mixture of oil and gas. Which sentences are incorrect about
the coating requirements for this valve?

A. Two-component epoxy is the correct coating system for such an
application.

B. The minimum surface cleanliness for applying the correct coating
on the substrate should be Sa 2.

C. The blasting should be done with chloride-free abrasives.

D. The correct coating system as per the NORSOK M-501 standard is
7A.

Answer: Option A is correct, as two-component epoxy is the right
coating choice. Option B is not correct, since as per NORSOK M501,
coating standard, the substrate cleanliness should be at least Sa 2½
according to ISO 8501-1. Option C is correct. Option D is not correct,
because NORSOK system 7A is used for components in splash zones
and not submerged areas. Therefore, options B and D are not correct.

10. What kind of information is included in coating data sheets?

A. Substrate material

B. Substrate roughness and coating thickness

C. Hazards involved in coating handling

D. Type of contaminants that can be removed by coating

Answer: Substrate material is not related to the coating data sheet,
so option A is not correct. Option B is not correct, because substrate
roughness and coating thickness are typically mentioned in the coating
specification and not the coating data sheet. Option C is correct, as the
hazards involved in using the coating are identified in the coating data

sheets. Option D is totally wrong, since coating is not used to remove surface contaminants.

BIBLIOGRAPHY

1. Andrew, W. (2008). *Handbook of Plastics Joining*, 2nd edition. Elsevier Science. Gulf Publishing, Amsterdam, Netherlands.
2. Bortak, T.M. (2002). *Guide to Protective Coating: Inspection and Maintenance.* United States department of the interior bureau of reclamation technical service center.
3. Det Norske Veritas (DNV) DNV-OS-D301 (2001). Fire protection. Hovik, Norway.
4. Equinor (2014). Technical and professional requirement TR 0042, surface preparation and protective coating. Equinor. 5th revision.
5. Goldschmidt, A. & Joachim, H. (2007). *Basics of Coating Technology*, 2nd edition. BASF Coatings AG, Münster/Germany.
6. International Organization of Standardization (ISO) 8501 (2007). Preparation of steel substrates before application of paints and related products—Visual assessment of surface cleanliness—Part 1: Rust grades and preparation grades of uncoated steel substrates and of steel substrates after overall removal of previous coating. Geneva, Switzerland.
7. International Organization of Standardization (ISO) 8502 (2020). Preparation of steel substrates before application of paints and related products—Tests for the assessment of surface cleanliness—Part 9: Field method for the conductometric determination of water-soluble salts. Geneva, Switzerland.
8. International Organization of Standardization (ISO) 8503 (2012). Preparation of steel substrates before application of paints and related products—Surface roughness characteristics of blast cleaned surface—Part 1: Specifications and definitions for ISO surface profile comparators for the assessment of abrasive blast cleaned surfaces. Geneva, Switzerland.
9. International Organization of Standardization (ISO) 8502 (2019). Preparation of steel substrates before application of paints and related products—Tests for the assessment of surface cleanliness—Part 3: Assessment of dust on steel surfaces prepared for painting (pressure–sensitive tape method). Geneva, Switzerland.
10. Ivanov, H. (2016). Corrosion protection systems in offshore structures. The University of Akron.
11. Jotun Marine Coatings (2001). Jotun Paint School. Sandefjord.
12. JOTUN (2021). Offshore installations. [online] available at: http://www.jotun.com/ww/en/b2b/paintsandcoatings/offshore-installations/ [access date: 7th March 2021].
13. Morales, R., et al. (2017). Challenging the performance myth of inorganic zinc rich vs organic zinc rich primers and activated zinc primers. NACE paper # NACE-2017-9127.
14. NORSOK M-501. (2012). *Surface Preparation and Protective Coating*, 6th edition. Elsevier, Lysaker, Norway.
15. Pugh, S., et al. (2003). Fouling during the use of seawater as coolant: The development of a user guide. *Engineering Conferences International*, Santa Fé, USA.
16. Prochaska, S. & Tordonato, D. (2017). Review of corrosion inhibiting mechanisms in coatings. Research and development office science and technology program. Report # ST-2017-1703.
17. Rasmussen, S.N. (2004). Corrosion protection of offshore structures. Hemple A/S.

18. Sotoodeh, K. (2018). Valve failures, analysis and solutions. *Valve World Magazine*, Vol. 23, No. 11, pp. 48–52.
19. Wang, J. (2020). Electrochemical investigation of corrosion behavior of epoxy modified silicate zinc rich coatings in 3.5% NaCl solution. *MDPI Journal of Coating*, Vol. 10, p. 444. doi: 10.3390/coatings10050444.

4 Coating Defects and Inspection

4.1 INTRODUCTION

As discussed in previous chapters, coating is used for different applications, and the most important among them is corrosion protection. All coating will fail and break down eventually over time. However, the service life of coating can be extended by performing maintenance painting when a coating defect or deterioration is identified. Dangerous coating failures occur before the predicted service life ends for one of the following reasons:

- Poor surface preparation of the substrate or metal surface before coating application; surface preparation of metal before coating application is explained in detail in Chapter 2;
- Poor coating application procedure;
- Poor-quality raw material;
- Poor coating system selection; more information about coating system selection is explained in Chapter 3;
- Improper coating formulation.

Coating failure has severe negative impacts, including financial loss, substrate mechanical or corrosion damage, costly rework costs and downtime. When coating failure occurs, the substrate is exposed to the environment until the defect is detected. Unprotected substrate could be corroded and be at risk of metal loss. The integrity of the structure could be jeopardized if the coating defect remains undetected for a long period of time. It is important to consider the cost of substrate repair due to coating failure. Due to the significance of coating failure, which is costly, time-consuming and potentially disastrous, this chapter is dedicated to addressing coating defects and failures, as well as coating inspection and procedures that may be taken to prevent them.

4.2 COATING FORMULATION

Coating formulation involves different concepts, such as consideration of the different ingredients in the coating, i.e., binders, pigments, additives and most importantly solvents. Solvents are essential formulation ingredients for various coating systems and applications. The reason why solvents are important is that coatings cannot perform well without them. Solvents affect important coating properties such as curing and viscosity. The other important formulation concept

DOI: 10.1201/9781003255918-4

is the *ratio of pigment to binder*. The pigment-to-binder ratio is calculated by the ratio of pigment mass to binder mass. Another important parameter is *pigment volume concentration (PVC)*, calculated according to Equation 4.1. PVC affects the gloss or shininess of the coating; a PVC of zero indicates zero pigment and highest shininess or gloss of the paint.

Pigment volume concentration (PVC) calculation

$$PVC = \frac{\text{Pigment volume}}{\text{Pigment volume} + \text{binder volume}} \qquad (4.1)$$

Other important concepts of formulation include density, *weight solids* and *volume solids*. Volume solids and weight solids should not be confused. Volume solids predict how much area a paint will cover. Weight solids indicate the weight of non-volatile ingredients in the paint. In simpler terms, volume solids and weight solids are the volume and weight of what is left on a metal surface after the paint dries. Equation 4.2 shows the relationship between the coating's wet-film thickness (WFT) and dry-film thickness (DFT). WFT refers to the thickness of the coating upon application before the coating dries and hardens. DFT refers to the coating thickness after it has dried.

Relationship between wet and dry-film thicknesses

$$DFT = \frac{WFT \times \%VS}{\%100} \qquad (4.2)$$

where:
DFT: Dry-film thickness (μm)
WFT: Wet-film thickness (μm)
%VS: Percent volume solids
Coating is made of volume solids and solvent; volume solids remain on the metal surface and form the dry coating film thickness. Solvent, on the other hand, is evaporated from the metal surface and wet coating film.

Example 4.1

The dry coating film is 100 μm. What is the WFT if the volume solids are 70%? What would be the percentage of solvents in the coating?

Answer

$$DFT = \frac{WFT \times \%VS}{\%100}$$

$$100 = \frac{WFT \times \%70}{\%100} \qquad WFT = 142.86\,\mu m$$

Percentage of solvent $= 100\% - \%$ volume solids$(VS) = 100\% - 70\% = 30\%$

The relationship between the dry and wet coating film is changed if thinner is added to the coating, as per Equation 4.3. Paint thinner is a kind of solvent that dissolves in the coating to reduce the viscosity of the paint.

Relationship between WFT and DFT by adding thinner to the coating

$$DFT = \frac{WFT \times \%VS}{\left(\%100 + \% \text{ added thinner}\right)} \tag{4.3}$$

Example 4.2

One liter of a coating with 70% VS is thinned 20%. What would the WFT need to be to get 100 μm of DFT eventually? What would the new value of VS percentage be after adding thinner to the coating?

Answer

Equation 4.3 is used to calculate the DFT after adding thinner.

$$DFT = \frac{WFT \times \%VS}{\left(\%100 + \% \text{ added thinner}\right)}$$

$$DFT = \frac{WFT \times \%VS}{\left(\%100 + \% \text{ added thinner}\right)}$$

$$100 = \frac{WFT \times \%70}{\left(\%100 + \%20\right)} \quad WFT = \frac{100 \times \%120}{\%70} = 171.4\,\mu m$$

A WFT of 171.4 μm is required when adding 20% thinner to achieve a final DFT of 100 μm.

The VS percentage would be changed after adding the thinner. The VS before adding thinner is 70%. Considering 1 L of coating, the volume of solids is 0.7 L. Now the volume of coating is increased by 20% from 1 to 1.2 L because of the added thinner. The percentage of solids in the thinned coating is the ratio of 0.7/1.2 equal to 58.33%.

The other important coating calculation is related to the theoretical volume of coating consumption (in liters), which is calculated through Equation 4.4.

Theoretical pint volume consumption in liters

$$PCV = \frac{A \times DFT}{10 \times VS} \tag{4.4}$$

where:

PCV: Paint consumption volume (liter);

A: Area of coating (m^2);

DFT: Dry-film thickness (micron);

VS: Percent volume solids

It is important to know that the PCV calculation provided in Equation 4.4 is based on the assumption that there is no loss of coating during the coating application. If a loss factor of 10% of the coating during application is considered, then 10% extra to the calculated PCV according to Equation 4.4 should be ordered to compensate for the loss factor.

Example 4.3

The coated area of a facility is 400 m^2. The WFT of the coating is 220 μm, and the VS% is 90%. Calculate the required volume of the paint, assuming a paint loss factor equal to 10%.

Answer

The first step is to calculate the DFT by knowing both the WFT and the VS% and using Equation 4.2.

$$DFT = \frac{WFT \times \%VS}{\%100} = \frac{220 \times 90\%}{100\%} = 198 \, \mu m$$

Now it is possible to calculate the PCV (theoretical paint volume consumption) by knowing the value of the DFT and using Equation 4.4.

$$PCV = \frac{A \times DFT}{10 \times VS} = \frac{400 \times 198}{10 \times 90} = 88 \, L$$

A PCV equal to 88 liters is the volume of the coating, considering no coating loss during implementation. However, since a 10% coating loss should be considered in this case, then 10% over the calculated coating volume should be ordered. Thus, the PCV considering a 10% loss is equal to 88 + 8.8 = 96.8 L.

4.3 COATING DEFECTS

4.3.1 RUN AND SAGS

Runs, sags and sagging refer to the downward "dropping" movement of the paint film immediately after application that causes uneven coating, meaning that some areas are thicker than other areas, as illustrated in Figure 4.1. Different factors could lead to sags and runs, such as excessively thick film, too much thinner, not

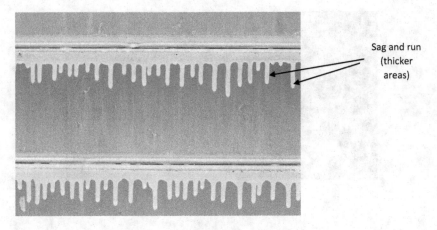

Sag and run
(thicker
areas)

FIGURE 4.1   Paint sagging. (Courtesy: Shutterstock.)

allowing the first coat to dry before applying the topcoat or a surface that is too hard or too glossy to keep the paint. The remedy actions for sagging are divided into two different conditions: before curing and after curing. If runs and sags happen before curing, the excess paint should be smoothened with a paintbrush and the paint spray condition should be altered. If the problem remains, the coating should be removed by sandblasting and the substrate should be prepared for new layers of coating.

4.3.2   ORANGE PEEL

This defect gives the paint finish a rough appearance similar to the outside skin of an orange. Figure 4.2 illustrates an "orange peel" coating failure in which the coating surface is like the skin of an orange. The main causes of this type of coating defect are poor application technique or wrong solvent selection, which results in rapid solvent evaporation. There are two remedy actions: one is to use the manufacturer's recommended thinner, and the other is to adjust the spray gun correctly. If the problem persists, then the solution is to remove the paint completely with sandblasting and prepare the substrate for a new coating.

4.3.3   BLISTERING

Blistering refers to the presence of small or large irregular bubbles in the coating, as illustrated in Figure 4.3. The bubbles could be broken or unbroken. Blistering can happen close to welded areas due to welding fumes remaining on the steel. The main cause of blistering is poor surface preparation, particularly the presence of contaminants on the metal surface. The main surface contaminant that causes blistering is soluble salt. Grease and oil can also cause coating blistering as well. The most important preventive action is to apply proper surface preparation.

FIGURE 4.2   Orange-peel coating failure. (Courtesy: Shutterstock.)

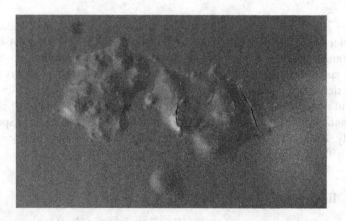

FIGURE 4.3   Coating blistering. (Courtesy: Shutterstock.)

Soluble salt testing should be performed according to the correct standards; ISO 8502-6 is used to extract soluble salt for analysis and 8502-9 can be used to measure the amount of soluble salt on a metal surface.

4.3.4 Blush (Blushing)

Blushing refers to white, milk-like deposits or haziness appearing on the coating film. Different factors could cause blushing, such as high humidity and cold or hot temperature during the coating application. The haziness or white blushing areas are in fact the participation of the polymers. Adjusting the temperature during coating application and avoiding coating application in high humidity are some of the preventive actions that can be taken to avoid blushing defects. Figure 4.4 illustrates a coating blush defect.

FIGURE 4.4 Coating blushing. (Courtesy: Shutterstock.)

FIGURE 4.5 Coating cratering. (Courtesy: Shutterstock.)

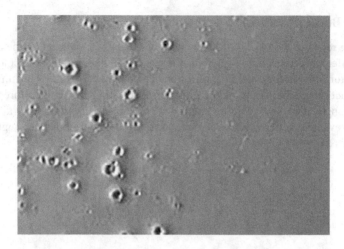

FIGURE 4.6   Fish-eyes defect. (Courtesy: Shutterstock.)

4.3.5 CRATERING (PITTING)

Cratering, also called pitting, is a kind of paint film defect in which small, rounded depressions are created in the coating, as illustrated in Figure 4.5. Cratering or pitting can be caused by different factors, such as air trapped inside the coating or surface contamination. Applying sandblasting, preparing proper surface and applying a new coat are remedies for this type of defect.

4.3.6 FISH EYES

This type of coating defect involves the separation or pulling apart of the wet film to expose the underlying finish or substrate. Fish-eye defect is very similar to cratering, as fish-eye pits open shortly after coating application (see Figure 4.6). Surface contaminants, such as oil, dirt or silicon, can cause this type of defect. As with cratering, the defective coating should be removed and the surface should be prepared for applying a new coating.

4.3.7 CHALKING

Chalking refers to the powdery deposit that appears on the surface of a paint due to the exposure of the paint to sunlight and ultraviolet (UV) light. Chalking can reduce the surface gloss and appearance of the paint. In fact, long exposure of coating to sunlight causes the pigments to become loose and less bound to the surface. Different types of pigments have different levels of sensitivity to UV light. As an example, natural pigments have less resistance compared to inorganic pigment types. Thus, epoxy coatings like zinc-rich epoxy (coating system 1B according to the NOROSK M-510 standard) and phenolic epoxy (coating system 6C as per the NORSOK M-510 standard) are not resistant to UV light. Figure 4.7

FIGURE 4.7   Chalking powders from the coating surface. (Courtesy: Shutterstock.)

FIGURE 4.8   Fading paint on a car. (Courtesy: Shutterstock.)

illustrates the hands of a person who has touched a coating surface with a chalking defect.

4.3.8   FADING

Fading is the gradual loss of the paint color; it is caused by the effects of sunlight and UV light. Chalking and fading can be connected to each other, meaning that chalking and fading can happen at the same time. However, chalking powders may hide the fading of the coating. As with chalking, organic pigments are more at risk of fading compared to inorganic coating. The other factor that could affect the vulnerability of the pigments to chalking and fading is cost. Typically, more expensive pigments have higher quality and are less prone to UV-related defects such as fading and chalking. Figure 4.8 illustrates fading paint on the right side of a car.

FIGURE 4.9   Coating dry spray defect. (Courtesy: Shutterstock.)

4.3.9   DRY SPRAY

Dry spray is a coating surface defect characterized by lack of brightness and a rough, dry, sandpaper-like texture, as illustrated in Figure 4.9. The problem occurs when the coating substance dries too quickly after excessive solvent evaporation. Different factors could cause dry spray, such as excessive paint viscosity, which could be due to incorrect or low-quality thinner. Poor spraying technique, caused by a dirty spray gun or incorrect compressed air pressure, could be other causes of dry spray.

4.3.10   WRINKLING

Wrinkling, also called lifting, refers to a paint surface texture with uneven waves. Wrinkling occurs because of warm weather, which causes coating to dry very fast. In fact, the coating surface skins over uncured paint not only because of warm weather but for other reasons, such as very thick coating. If the coating is too thick, the surface paint can dry while the under paint remains wet. Figure 4.10 illustrates an orange paint with signs of wrinkling. The remedy action is to scrape off the wrinkles and apply a thinner coat. In addition, intense sunlight should be avoided during coating application.

4.3.11   CRINKLING

Crinkling refers to the swelling and lifting of the substrate layers when new paint is applied. This problem can occur during the painting process or during drying. Figure 4.11 illustrates a crinkling defect in a blue coating. There are different causes of crinkling. One is to ignore the flash-off time and apply a fresh coating layer over a wet lower layer. Flash-off time is the necessary

FIGURE 4.10 Wrinkling defect. (Courtesy: Shutterstock.)

FIGURE 4.11 Crinkling defect. (Courtesy: Shutterstock.)

waiting time taken to recoat or spray once the first coat has been applied to the substrate. The other factor that could cause crinkling is that the first coating layer applied is not thick enough. The third cause could be that the substrate has not properly dried. Recoating a substrate with the wrong repair coating material is another factor that can cause crinkling. The solution to prevent crinkling defect is to comply with the recommended flash-off time. Using the correct coating film thickness is the other important consideration to prevent this type of defect. The film thickness should be correct according to the process data sheet. The last important prevention measure is to ensure that the substrate is dry.

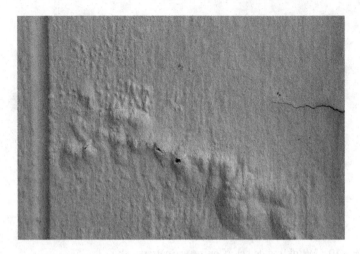

FIGURE 4.12   Swelling defect. (Courtesy: Shutterstock.)

FIGURE 4.13   Solvent boil painting defect. (Courtesy: Shutterstock.)

4.3.12 SWELLING

Swelling refers to the very slow evaporation of residual solvent from a freshly coated paint process. Slow evaporation causes both swelling and sweating of the paint surface, which can be widespread. The other factor that could cause this type of paint defect is an excessive amount of moisture on the substrate, which causes the coating film to swell and lose its adhesion to the substrate. Too short a drying time and too great a film thickness are other causes of swelling defect. Figure 4.12 illustrates the ingress of water beneath a coating layer, which results in coating swelling. Coating swelling, as illustrated in the figure, is very similar to coating blistering. Different actions can be taken to prevent swelling defects, such as ensuring the recommended film thickness and required drying time.

4.3.13 Solvent Boil

A paint surface may become blistered because of solvent entrapment in the surface of the paint film. Solvent boil can occur when the applied coating is too thick, the flash-off time between individual paint coats is too short or when wet coating is applied over another wet layer. The other cause of this type of defect is selecting the wrong hardener or thinner, which acts too quickly and prevents the evaporation of the solvent from the coating. The other cause is excessive coating thickness. The appearance of solvent boil is very similar to fish eye and pitting, as illustrated in Figure 4.13.

4.3.14 Pinhole and Holiday

Pinholes refer to small holes or craters in fresh paint. If solvent boil defects are not removed, they can result in pinhole marks. Early inspection to detect the pinholes, and applying additional coats after mechanical or blast cleaning can prevent pinhole coating defect. A "holiday" refers to defects inside the coating, mainly in the form of holes, like pitting, pinholes, etc. A holiday detector can be used for the inspection of possible coating defects. Figure 4.14 illustrates pinhole defect in the applied coating system.

4.3.15 Delamination

Coating delamination refers to paint peeling off the substrate or undercoat. Delamination occurs due to loss of coating adhesion and the subsequent separation and lifting of the paint from the substrate, or a poorly bonded undercoat. The

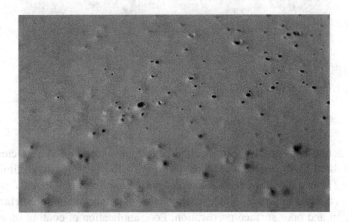

FIGURE 4.14 Pinhole defect. (Courtesy: Shutterstock.)

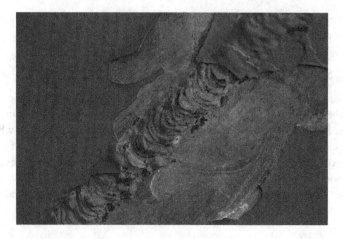

FIGURE 4.15   Delamination. (Courtesy: Shutterstock.)

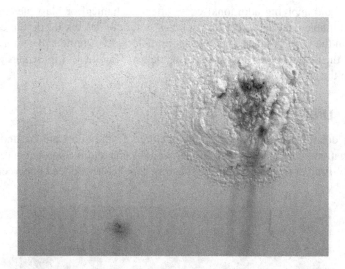

FIGURE 4.16   Undercutting. (Courtesy: Shutterstock.)

remedy action is to sandblast or mechanically remove the loose paint, clean and roughen the smooth surface and recoat it. Figure 4.15 illustrates the delamination of coating from a substrate. Delamination destroys coating strength and durability as well as coating appearance. The main causes of delamination are surface contamination and poor surface preparation. Poor application of coating on a metal surface is another cause of delamination.

FIGURE 4.17   Coating cracking. (Courtesy: Shutterstock.)

4.3.16 UNDERCUTTING

Undercutting refers to blistering or peeling of paint in which the substrate is exposed to a corrosive environment and becomes rusted. In fact, corrosion progresses under the coating in this kind of coating defect. Figure 4.16 illustrates paint blistering and the rust and corrosion of the substrate known as undercut. The rust and corrosion products seen in the picture lift the paint from the metal surface, such that the corrosion and rust will grow worse over time if not remedied. Two mitigation actions can be taken to prevent this type of coating defect: early detection of undercutting with a holiday detector and the use of inhibitive pigments in the coating to prevent corrosion attack.

4.3.17 CRACKING

This type of coating defect and failure refers to the deep cracks in the coating that result in the exposure of the substrate to the environment. There are different causes of cracking, such as paint shrinkage, limited painting flexibility, applying excessive thick coating or applying and/or curing the coating at a very high temperature. The solution is to remove the old, cracked coating, apply sandblasting on the metal surface and then apply a new coating. Figure 4.17 illustrates cracking.

4.3.18 BLEEDING

Paint bleeding refers to the staining or seeping through a pigment from the substrate into the topcoat. Bleeding is normally observed as spots or patches of discoloration on the topcoat, often in red or yellow color. One of the causes of bleeding is when a repair coat is applied on a surface that has not been properly cleaned of old paint. In such a case, pigments from the old paint dissolve in the

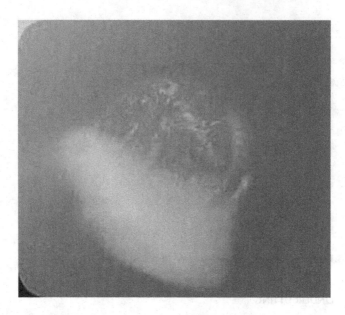

FIGURE 4.18   Bleeding coating defect. (Courtesy: Shutterstock.)

solvent of the repair coating and cause shading on the repair coating surface. The solvent of the new coating can be tested to make sure that it is free from pigments. In addition, it is important to clean the surface properly of old paint before applying a new coating. Bleeding is common in cases when a light repair coat is applied over old, dark paint. Figure 4.18 illustrates bleeding coating defect.

4.4   COATING INSPECTION

4.4.1   ESSENCE OF COATING INSPECTION AND INSPECTOR

Coating a component entails the investment of some additional time and money on the component beyond the cost of the coating materials. Of course, it is essential to have a very well-written coating specification prepared by the owner or client, i.e., the party that owns and operates the plant where the component is located. Even a very accurate and well-written coating specification cannot prevent significant coating defects, since there is no guarantee that the requirements of the specification will be met without proper inspection. Lack of or improper coating inspection has negative consequences, such as costly maintenance to the damaged coating; coating defects and corrosion of the metal surface; health, safety and environment (HSE) problems; low quality of the final job; costly delay in the implementation of the job, etc. A coating inspector imposes some additional costs on the project, related to time, travel and inspecting materials. Depending on the size of the project, the inspector cost could be 5%–18% of the total cost of the coating. But the value of the inspection and inspector is typically higher than the

FIGURE 4.19 Coating inspectors inspecting the coating on a ship hull. (Courtesy: Shutterstock.)

cost since the presence of the inspector is a kind of insurance for the client and decreases the risk of early and catastrophic coating failure.

To achieve a high-quality coating application that meets the relevant standards, two main parameters should be considered. One is quality assurance, which includes all the documents and management system for coating activities. The second is coating inspection and testing. Coating inspection is a very important task in coating applications to ensure the quality of the coating and prevent coating defects. The tasks of an inspector can be summarized as follows: ensuring that the requirements of the coating specification are met, verifying the quality of the work that is done by the valve and actuator supplier or in the construction yard, carrying out the necessary inspections, preparing the written records of the coating job and issuing reports. The inspector should have a good knowledge and overview of the relevant coating specifications and standards. In addition, the inspector should be skillful in methods of inspection and inspection tools, and the methods involved in substrate cleaning, surface preparation, etc. Figure 4.19 illustrates coating inspectors inspecting the final coating on the hull of a ship and preparing a report. A paint inspector must deal with several parties, such as the customer or owner, the construction yard personnel and the supplier of the paint and painting components.

4.4.2 COATING CHECKPOINTS

Coating inspection is not limited to the inspection and control of the coating itself after its application on the substrate. The inspection should be performed at different critical phases, such as steelwork, pre-surface treatment, after surface preparation and before coating application and during and after coating application. It is essential that the inspector verify and document that all the performed

FIGURE 4.20 Grinding welding defects from a metal surface. (Courtesy: Shutterstock.)

FIGURE 4.21 Measuring surface roughness of steel. (Courtesy: Shutterstock.)

activities, such as steelwork and surface preparation, are done in accordance with the project requirements and applicable standards.

Steelwork is explained more in detail in Chapter 2. In short, steelwork is carried out to remove edges, grind off welding defects and repair metal surface defects. The complete definition and list of such defects are given in ISO 8501-3. The ISO

12944-3 standard provides the acceptance level of various surface defects. Some of the most important steelwork activities are rounding the edges, grinding the welds, removing weld defects and grinding the laminations. Figure 4.20 illustrates a worker grinding welding spatter defects from a steel surface.

The other inspection stage involves checking the metal surface to ensure that it is free of oil, grease and salt prior to sandblasting. Surface salt analysis and examination are addressed in ISO 8501-6 and 9. Two main standards used to assess surface cleanliness are ISO 8501-1 and ISO 8501-2. The second standard addresses the surface cleanliness for repair coating, meaning that some coating remains on the substrate before surface preparation for a new coating. Other checking that could be done, especially after sandblasting, is to measure the dust according to ISO 8502-3. Surface roughness is another important factor in surface preparation before coating application, which should be in compliance with ISO 8503. Figure 4.21 illustrates an inspector measuring the surface roughness of a steel surface.

Humidity is another parameter that can cause coating defects. Humidity and other weather conditions affect the speed and intensity of the reactions involved in paint curing and protective film formation. The maximum weather humidity in many coating specifications ranges between 50% and 85%. Coating manufacturers may also have some advice regarding the relative humidity during the coating application. If the air and surrounding environment are too dry, the paint can cure too quickly, which will cause cracks and other damage. Humid air can prevent the evaporation of solvent from the coating and jeopardize coating curing and can cause other types of coating damage like bubbles and blisters. Thus, it is important that the coating inspector measure the relative humidity, air and steel temperature and the dew point.

The dew point is important; if the substrate's surface temperature reaches the dew point temperature, condensation can form, which puts the metal surface at risk of oxidation and rust occurrence. Like the effect of moisture on the coating, condensation that forms at the dew point will cause the coating not to dry or cure properly. Both the steel and air temperature have a direct impact on condensation formation and dew point, so they should be checked by the coating inspector as well.

The inspector's task during coating application is to ensure that all the coating tasks are performed according to the project specification and the applicable standards. The inspection covers all the work that is done, from opening the paint tin to the application of the final coating. The inspector should perform the following tasks:

• Checking the coating data sheet in order to identify the correct thinner that should be used for the coating, and making sure of the thinner application.
• Checking the right curing agent for the base coating. A curing agent is a substance used with coating in order to harden the coating after application and facilitate curing.

- For the application of some coatings, like two-component epoxy in which two components should be mixed to form the final coating product, the inspector should make sure of the proper mixing of the two components and the quality of the final product.
- Ensuring that the number of coatings is applied according to the project specification and standard.
- Checking the coating primer and the dry and wet thickness of the coating.
- Checking the atmospheric conditions during coating implementation; it is also important to make sure that the area where the coating is implemented is ventilated properly for health, safety and environmental reasons.
- Each type of paint requires a specific amount of time to dry and harden, which is called curing time. The curing time should be given in project documents such as the coating data sheets. The inspector should make sure that the coating is allowed to dry for the correct amount of time.
- Checking the coating DFT; it is important for the inspector to make sure that the minimum required coating thickness according to the coating specification and standards are met over the whole length of the coating.

4.4.3 INSPECTOR CAPABILITY LEVELS

Coating inspectors are categorized according to their experience and education into different levels. Inspectors are very commonly categorized into three levels: I, II or III. Coating inspectors should have enough experience, knowledge and education to competently perform the inspection.

A level I inspector has the lowest requirements for education, training and experience. A level I inspector should be at least a high-school graduate and have at least 6 months of experience in equivalent inspection activities. As an alternative, a level I inspector could have a college degree or higher plus 3 months of work experience in coating inspection activities. The responsibilities and capabilities of a level I inspector are listed as follows:

- Implement and record all the inspections applicable according to the required procedures;
- Verify the coating tools and instrument calibration;
- Perform inspections in accordance with the applicable procedures.

A level II inspector has a medium level of requirements for education, training and experience. A level II inspector should meet at least one of the following requirements: (1) high-school graduation plus 1 year of satisfactory performance as a level I coating inspector; (2) completion of college-level work plus 1 year of experience as a coating inspector; (3) a 4-year college degree plus 6 months of coating inspection experience. The roles, responsibilities and capabilities of a level II inspector are summarized as follows:

- Perform all the duties and responsibilities of a coating inspector level I;
- Plan and supervise the inspections, initiate and review the inspection procedures and evaluate the sufficiency of the activities;
- Revise, organize and approve the results of inspection;
- Monitor the performance and supervise the work of level I coating inspectors;
- Train and verify the qualifications of level I coating inspectors and issue the relevant certificates;
- Initiate some changes for improving the quality of coating procedures;
- Implement quality assurance programs.

A level III inspector has the highest level of requirements for education, training and experience with regard to coating issues. A level III coating inspector should, at minimum, meet at least one of the following requirements: (1) high-school graduation plus 6 years of satisfactory performance as a Level II coating inspector; (2) high-school graduation plus 10 years of inspection activities; (3) high-school graduation plus 8 years of equivalent inspection activities with at least 2 years of experience as a Level II coating inspector; (4) completion of college-level work with an Associates degree and 7 years of experience in inspection activities with at least 2 years of this experience in industrial facilities using high-technology coating; (5) 4-year college degree plus 5 years of experience in equivalent inspection activities with at least 2 years of this experience associated with industrial facilities using high-technology coating. The roles and responsibilities of a level III inspector are summarized as follows:

- Carry out all the duties and responsibilities of a level II inspector;

TABLE 4.1
Sample of Coating ITP

Coating Inspection Test Plan Task	Coating Manufacturer	Contractor Inspector	Client
Steelwork, including the preparation of the metal surface by removing weld defects, sharp edges and surface defects as per ISO 8501-3	H	W	R
Soluble salt measurement according to ISO 8502-6	H	W	R
Oil and grease removal by solvent cleaning as per SSPC-SP 1	H	W	R
Sandblasting of the metal according to ISO 8501-1 Sa 2½	H	H	R
Assessment of the surface dust as per ISO 8502-3	H	W	R
Surface profile measurement as per ISO 8503	H	H	R
Dry-film thickness measurement	H	H	R

- Perform level, I, II and III coating inspection tasks;
- Evaluate programs for training coating inspectors;
- Authorize level II inspectors to carry out coating training programs;
- Approve all safety-related coating procedures.

4.4.4 INSPECTION AND TEST PLAN (ITP)

Typically, an ITP or quality control plan is generated in every project to include a detailed description of the testing and inspection that should be performed on the equipment and facilities. Similarly, an ITP is generated for coating inspection by the coating manufacturer and submitted to the contractor company and owner (client) for review and approval. Table 4.1 shows a small sample of an ITP indicating the type of tests that should be performed on the coating.

Coating inspection tasks are listed in the left column of the table, and the three columns on the right are allocated to the level of inspection authority of the three parties: the coating manufacturer, the inspector (typically from a contractor company) and the client or owner. The letters H, W and R indicate the level of authority for each party with regard to coating inspection; they are defined as follows:

> *H (hold point):* Hold point indicates that the relevant inspection activity cannot proceed or be concluded without the approval of the designated authority. In fact, the work cannot proceed until the receipt of a hold point release is issued by the authorized person or party. As an example, the hold point is allocated to the inspector for sandblasting of the metal. This means that the coating supplier cannot move on to the next task after sandblasting, which is an assessment of surface dust, before getting approval from the contactor regarding the sandblasting activity.
>
> *W (witness point):* Witness point means that the relevant party must physically attend during the activity or test. However, if the party that can

FIGURE 4.22 Mirror used to inspect a surface visually. (Courtesy: Shutterstock.)

FIGURE 4.23   Welding visual inspection with a magnifier. (Courtesy: Shutterstock.)

witness the test cannot attend, she/he can send a notification allowing the test or task to proceed without her/his presence.

R (review): Review means that the inspection records are sent to the relevant party (e.g., the client) for review.

4.4.5   INSPECTION TOOLS AND METHODS

4.4.5.1   Visual Inspection

The first group of inspection tools are used for visual inspection. Visual inspection tools typically include a mirror, flashlight and magnifier. These handy tools enable the inspector to detect defects visually and verify the cleanliness and roughness of the surface. Figure 4.22 illustrates a mirror that is used to check the surface and identify any welding defects on the metal surface prior to applying coating. Figure 4.23 shows a visual inspection of welding using a magnifier. Some welding defects, like spatter and porosity, should be removed from the metal surface before applying the coating.

The other simple tool that is used for marking areas with a defect is chalk. Chalk can be used for marking after surface preparation, before coating and after coating. Any areas that have defects should be marked properly with chalk. Before using a chalk, it is important to make sure that it will not have any negative effect on the applied coating. Chalk can have contaminative effects on some kinds of coatings.

Assessment of surface cleanliness is performed according to the different parts of ISO 8502, summarized as follows:

Part 1: Field test for soluble iron corrosion products;

Part 2: Laboratory determination of chloride on cleaned surfaces;

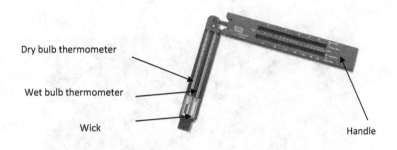

Dry bulb thermometer

Wet bulb thermometer

Wick

Handle

FIGURE 4.24 Sling psychrometer for humidity measurement. (Photograph by author.)

Part 3: Assessment of dust on steel surfaces (pressure-sensitive tape method);

Part 4: Guideline on the estimation of the probability of condensation prior to paint application;

Part 5: Measurement of the chloride on steel surfaces prepared for painting (ion detector tube method);

Part 6: Extraction of soluble contaminations for analysis (the Bresle method);

Part 9: Conductometric measurements of soluble salts.

Note: Parts 7, 8 and 10 of ISO 8502 are not prepared yet.

The other simple tool that is used to estimate the amount of dust on the metal surface as a part of surface cleanliness examination and inspection is tape. More information about dust assessment on a metal surface is available in Chapter 2.

4.4.5.2 Weather Conditions Measurement

A sling psychrometer (see Figure 4.24) can be used to determine the relative humidity, which is expressed as a percentage, and to measure the vicinity temperature. The reasons why measuring the humidity is important for coating application were mentioned earlier in this chapter.

Main advantages of this tool are that it is cheap and does not require any power source. In addition, the device is portable and requires little maintenance. A sling psychrometer has two thermometers: a dry bulb and a wet-bulb thermometer. Both are typically enclosed in a plastic housing. In addition, this device has a handle and a wick, which are highlighted in Figure 4.24. The wick covers the bulb of the wet-bulb thermometer; it should be soaked in water before using the tool for measurement. The working principle of a sling psychrometer is based on water evaporation from the surface. When water on the surface evaporates, it extracts heat, which in turn results in the surface cooling

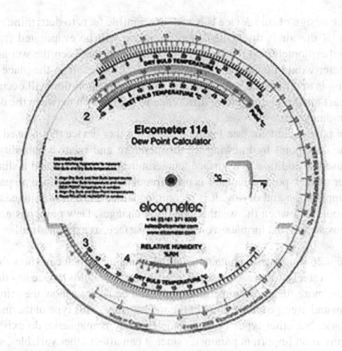

FIGURE 4.25 Dew-point calculator. (Photograph by author.)

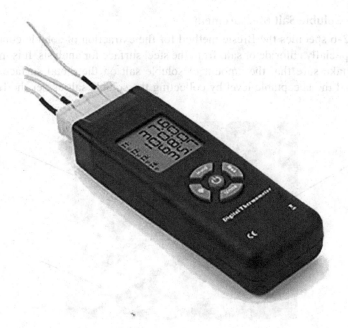

FIGURE 4.26 A metal surface thermometer. (Photograph by author.)

down. The design of this device is based on scientific facts to determine relative humidity. If the air is drier, then more moisture will be evaporated from the wet-bulb thermometer, so the temperature difference between the wet and dry thermometer would be higher. On the other hand, if the air in the place where the coating is to take place is more humid, then less evaporation will occur from the wet part and less temperature difference will be shown between the dry and wet thermometer.

A dew point calculator (see Figure 4.25) is another device that is used to calculate the dew point by knowing the temperature and relative humidity. Dew point is used in addition to a surface temperature thermometer and a sling psychrometer. A dew-point calculator is used frequently during surface preparation, coating application and drying. It is recommended to use a dew-point calculator every 6 hours and when the weather condition changes. Dew point has an effect on condensate and rust formation on the metal surface, as explained earlier in this chapter.

Figure 4.26 illustrates a contact metal thermometer, which can show the temperature of a steel surface. As it can be seen in the figure, this device has the ability to show more than one temperature, so it can be used to show the minimum, maximum and average temperature of the steel surface. This type of thermometer is electronic, but other types of thermometer, such as magnetic, do exist. Steel temperature is an important parameter since it can affect other variables, such as the speed and extent of coating curing, the service life of the coating, the type of coating and the number of coating layers.

4.4.5.3　Soluble Salt Measurement

ISO 8502-6 specifies the Bresle method for the extraction of soluble contaminants, especially chloride or salt, from the steel surface for analysis. It is important to make sure that the amount of soluble salt on the metal surface does not exceed the acceptable level by collecting the soluble salt as per the Bresle

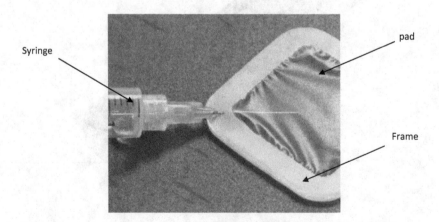

FIGURE 4.27　Bresle test performance. (Courtesy: Shutterstock.)

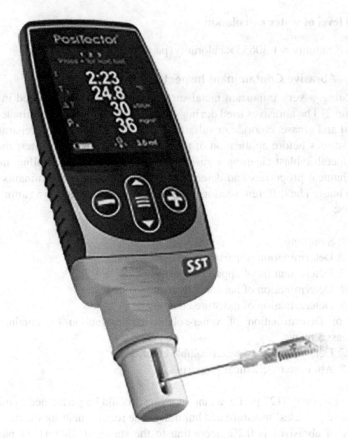

FIGURE 4.28 Salt measurement as per ISO 8502-9. (Photograph by author.)

method and then applying a salt test to prevent premature coating failure. The method used in the Bresle method is known as surface extraction. A volume of water is placed in a chamber against the surface. The water dissolves the soluble salts on the surface, which increases the electrical conductivity of the water. Thus, the measurement of the electrical conductivity of the water indicates the level of salt. The maximum content of soluble impurities on the surface shall not exceed 20 mg/m^2 according to the ISO 8502-9 standard. However, the project coating specification could indicate the acceptable amount of soluble salt. Figure 4.27 illustrates how a Bresle test is performed. The water is injected by a syringe through the frame and inside the pad. The water should be pumped in and out through the syringe several times according to the standard.

After getting different samples, the syringe is connected to a conductivity meter as per the ISO 8502-9 standard (see Figure 4.28). The salt level is measured by assessing the conductivity level of the solution. Higher amounts of salts and impurities in the water indicate higher conductivity. The amount of chloride affects the salinity of the water. The salinity level is calculated according to Equation 4.5.

Salinity level of water calculation

$$\text{Salinity} = 1.80655 \times \text{chlorinity} \left(\text{partsperthousand or g per kg}\right) \qquad (4.5)$$

4.4.5.4 Abrasive Contaminant Inspection

Sandblasting, a very important metal surface preparation, is discussed in detail in Chapter 2. The abrasives used during sandblasting could be contaminated with water, oil and grease, chloride or sulfates. ISO 11127, standard for preparation of steel substrates before application of paints and related products—test methods for non-metallic blast cleaning abrasives, is the reference for sampling, measuring mechanical properties and determining and evaluating contaminants in the abrasive blast. The different sections of the ISO 11127 standard are summarized as follows:

Part 1: Sampling;
Part 2: Determination of particle size distribution;
Part 3: Determination of apparent density;
Part 4: Determination of hardness by a glass slide test;
Part 5: Determination of moisture content;
Part 6: Determination of water-soluble contamination by conductive measurement;
Part 7: Determination of water-soluble chloride;
Part 8: Abrasive mechanical properties.

According to ISO 11127 part 5, a laboratory test should be performed to measure the abrasive particles' moisture and humidity. The requirement for the maximum humidity of abrasives is 0.2% according to the standard. ISO 11127 part 6 is used to measure the content of water-soluble contamination. A sample of 100 g of abrasives and 100 mL of water should be taken and mixed. The abrasive and

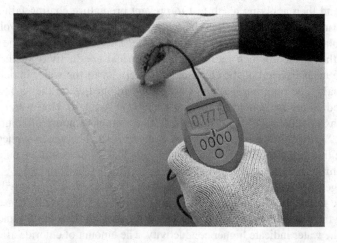

FIGURE 4.29 Electromagnetic-coating DFT measurement. (Courtesy: Shutterstock.)

water mixture should be mixed for 5 minutes and let rest for 1 hour. After the 1-hour rest, the mixture is required to be shaken for another 5 minutes. Then the conductivity of the solution is measured with a conductivity meter; it should be <25 milli siemens per meter. Siemens is the unit of measurement of electrical conductivity; it is exactly opposite to the electrical resistance of a substance. Electrical resistance, expressed in ohm (Ω), measures the opposition against the flow of the electrical current. Electrical conductivity measures how an electrical current moves within a substance. Siemens is expressed in Ω^{-1}.

The other important abrasive contaminants are oil and grease. The American Society for Testing and Materials (ASTM) D4940, standard for test method for conductimetric analysis of water-soluble ionic contamination of blast abrasives, is used to measure the grease and oil. The abrasives should be placed inside a container, and the container should be filled in with clean, fresh water. Then the mixture should be shaken, and the top of the mixture should be observed to see if oil and grease are present.

4.4.5.5   Coating DFT Measurement

DFT measurement is another type of inspection carried out after coating application and hardening. Many different methods can be used for this purpose. However, the magnetic or electromagnetic method is very common to determine coating film thickness, as illustrated in Figure 4.29. The main working principle of an electromagnetic coating thickness measurement device is based on an electronic instrument that uses an electrical circuit and magnetic induction to convert the reference signal to a coating thickness reading. The measurement of coating thickness should be performed more than once and on different coating spots for more accurate results.

4.4.5.6   Coating Adhesion Measurement

Coating adhesion evaluation is another important test to ensure the satisfactory performance and durability of coating over its design life. Three types of testing, knife, pull-off and tape tests are the most common methods for assessing coating adhesion. The knife test is the simplest test of coating adhesion; in this test, a knife is used to evaluate the adhesion of the coating to the substrate. In general, the degree of difficulty in removing the coating from the metal surface and the size of the removed coating chip are converted to the degree of coating adhesion. The well-known standard for coating adhesion test by knife is ASTM D6677, standard test method for evaluating adhesion by knife. Although this method is qualitative and the result of the test is largely dependent on inspector experience and interpretation, this method of testing has been widely used in the industry for many years. It is not possible to correlate the results of this test to other adhesion tests, such as pull-off, tape, etc. The knife test method can be both in the laboratory and in the field.

The knife adhesion test can be summarized in five simple steps:

1 Select the test area;

FIGURE 4.30   Coating adhesion test by knife. (Courtesy: Shutterstock.)

2 Make a single cut by knife through the coating in 1.5″ length. A ruler is
   proposed to be used to make the cut straight;
3 Make a second cut of the same length (1.5″);
4 While making the second cut, it should be kept in mind that the first
   and second cuts should form a sort of "X," and there should be 30°–45°
   between the two cuts;
5 Attempt to lift the coating with the tip of the knife.

Figure 4.30 illustrates a coating adhesion test by knife, highlighting the
X-shaped cut on the coating.

The more formal version of the adhesion test is the tape test. The cutting tool,
which could be a sharp razor blade, scalpel, knife or other fine-edged cutting
device, is used to cut the coating. Tape is applied over the incision in the coating
and then removed. The cuts made in the tape test are similar to those of the knife
test. There are two types of tape tests: one involves making an X-shaped cut (test
method A), and the other is the hatch tape test (test method B). The tape adhesion
test on coating is performed according to the ASTM D3359 standard. In short,
in test method A, an X-shaped cut is made through the film to the substrate,
pressure-sensitive tape is applied over the cut and then removed, and adhesion is
assessed qualitatively on a zero-to-five scale. Test method B is a lattice pattern;
either 6 or 11 cuts are made in each direction through the film to the substrate;
and pressure-sensitive tape is applied over the lattice pattern and then removed,
and adhesion is assessed qualitatively on a scale of 0–5. Test method A is pre-
liminary and intended for use in the field, while test method B is suitable for use
in the laboratory or shop environment. Test method B is not considered suitable
for coating film thicker than 125 μm equal to 5 mils, unless wider-spaced cuts

FIGURE 4.31   Hatch tape test (test method B). (Courtesy: Shutterstock.)

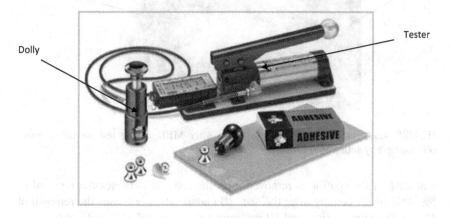

FIGURE 4.32   Pull-off test devices. (Photograph by author.)

are made, in which case there would be an agreement between the purchaser and seller. The tape used for this test should be 25 mm equal to 1″ wide and transparent with an adhesive peel strength of between 6.34 and 7.00 N/cm. For X-cut tape method A, the highest level of coating adhesion is 5A, and the lowest level is 0A. 5A level indicates no peeling or removal. 4A indicates trace peeling or removal along the incisions or at their intersection. 3A represents jagged removal along the incisions up to 1.6 mm on either side. 2A represents jagged removal along the incisions up to 3.2 mm on either side. 1A indicates removal from most of the area of the X under the tape, and 0A indicates removal beyond the area of the X. Figure 4.31 illustrates a hatch tape test (test method B). The 5B classification indicates that no coating is removed from the surface. 4B classification means

FIGURE 4.33 MEK rub test. (Photograph by author.)

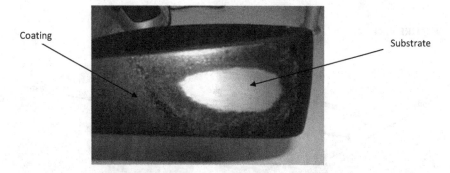

FIGURE 4.34 Penetration to the substrate after MEK curing test on the coating. (Photograph by author.)

that <5% of the coating is removed after the test. 3B indicates the removal of 5%–15% of the coating after the test. 2B classification indicates the removal of 15%–35% of the coating, and 1B indicates the removal of 35%–65% of the coating. 0B, the result of the test illustrated in Figure 4.31 shows the removal of more than 65% of the coating.

The pull-off test is the more accurate and quantitative test for adhesion compared to both the knife test and the tape test. A loading fixture, called a dolly or stub, is affixed to the coating. By using a portable pull-off adhesion tester, a load is applied constantly and increased until the dolly is pulled off the metal surface. The amount of force required to pull the dolly off can be measured in pounds per square meter (psi) or mega pascal. The tester can be operated manually, pneumatically or by hydraulic oil. Figure 4.32 illustrates pull-off test devices, including dolly and tester.

4.4.5.7 Coating Curing Test

There are numerous ways to test paint curing, but the most common and traditional way is to use methyl ethyl keton (MEK); this method is known as the MEK
rub test (see Figure 4.33). MEK is a highly volatile and flammable liquid solvent.
The MEK curing test for inorganic zinc-rich primers is performed according to
the ASTM D4752 requirements.

The procedure involves six steps; the first step is to select areas of the coating surface that are at least 150 mm long on which to run the test. Clean the
area with tap water and dry it with cloths afterward to remove all surface residue and contaminants. The second step is to measure the DFT of the coating in
different areas of the test and compare them with the coating specification to
make sure that the coating thickness is correct. The third step involves folding
a cloth and wetting it with the MEK solution. Step four involves placing the
MEK-wetted cloth on the test area and rubbing it by moving it back and forth
on the coating surface. One pass forward and back is considered a "double
rub," which should be performed in ~1 second. Step five involves continuing
to rub the coating until either the metal substrate is exposed or 50 double rubs
have been completed. The last step is to rate the result according to the following criteria.

Resistance rating 0 is the lowest curing level, indicating penetration to the substrate in 50 double rubs or less. Resistance rating 1 means a heavy depression in
the film but no actual penetration to the substrate after 50 double rubs. Resistance
rating 2 means heavy marring and obvious depression in the film coating after 50
double rubs. Resistance rating 3 means some depression in the film coating thickness after 50 double rubs. Resistance rating 4 means burnished appearance in
the rubbed areas and a slight amount of zinc on the rubbing cloth after 50 double
rubs, and finally resistance rating 5 means no amount of zinc on the rubbed cloth
after 50 double rubs. As an example, Figure 4.34 illustrates resistance rating 0,
meaning the lowest level of curing and penetration to the substrate in 50 double
rubs or less.

4.5 CONCLUSION

This chapter focuses mainly on the main coating defects. In addition, some of
the essential coating calculations and formulation are covered in this chapter.
Coating inspection techniques and tools and coating inspector qualifications are
other important topics covered here.

4.6 QUESTIONS AND ANSWERS

1. The WFT of a coating is 300 μm, and the solid volume percentage is
   66%. The area subject to coating is 500 m². Considering a paint loss factor of 20%, which sentence is correct?
   A. The DFT is equal to 210 μm.

B. The volume of the coating should be increased by 80% because of the loss factor of 20%.

C. Adding thinner to the coating increases the solid volume percentage.

D. The volume of coating consumption, considering the 20% loss factor, would be 180 L.

Answer: The DFT can be calculated by knowing the WFT and the percentage of volume solid (%VS) and applying them in Equation 4.1.

$$DFT = \frac{WFT \times \%VS}{\%100} = \frac{300 \times \%66}{\%100} = 198\,\mu m$$

Thus, option A is not correct, since the value of DFT is given equal to 210 μm. Option B is not correct, because the volume of the coating should be increased by 20% when the loss factor is 20%. Option C is not correct, since adding thinner to the coating reduces the solid volume percentage. The next step is to calculate the volume of coating consumption through Equation 4.4.

$$PCV = \frac{A \times DFT}{10 \times VS} = \frac{500 \times 198}{10 \times 66} = 150\ L$$

However, the loss of factor of the coating is 20%, so 20% of the calculated PCV should be added to the calculated PCV, meaning the total volume of coating consumption is $150 + (20\% \times 150 = 30) = 180\,L$. Thus, option D is the correct answer.

2. Which of the following types of coating defects have similarities with regard to their cause?

A. Fading and chalking

B. Running and blooming

C. Blistering and pitting

D. Options A and C are both correct.

Answer: Fading and chalking are both the outcome of the UV light effect on the pigments, so option A is correct. Blooming is similar to blushing, as the surface of the coating develops haziness after drying out. The cause of blooming is the movement of some pigments to the surface of the paint when it is drying. On the other hand, running can occur due to different factors, such as film that is too thick or too thin. Thus, option B is not correct. Option C is correct, as both pitting and blistering are caused by surface contaminants such as oil and grease. Since both options A and C are correct, the correct answer is D.

3. Which two types of coating effects create pits on the paint surface?

A. Dry spray and delamination

B. Fish eyes and cratering

C. Bleeding and orange peel

D. Wrinkling and holidays

Answer: Dry spray creates a rough, dry, sandpaper-like surface. Delamination is the separation or lifting of paint or the peeling of paint from the undercoat or substrate. Therefore, neither dray spray nor delamination creates a pit on the paint surface, so option A is not correct. Option B is correct, because fish eyes and crating are both characterized by pitting on the surface. Bleeding is the staining or discoloration of the applied paint by the previous coating. This defect typically occurs when a light-colored coat is applied over a dark-colored coat. Orange peel, like dry spray, provides a very rough surface and appearance. The rough surface of the orange peel defect is like the skin of an orange. Thus, option C is not correct, because bleeding and orange peel do not create pits on the coating surface. Wrinkling causes wavy lines that appear in the paint film during coating application. Holiday refers to holes, which are like pits, so option D is partially correct. In conclusion, option B is the correct answer.

4. Fill in the gaps with the correct words.

Moisture is trapped under the coating, and wet-to-wet coating systems have been applied on top of each other. For these two reasons, _____ defect is observed in the coating, and it has loosened its adhesion to the substrate. In another case, the coating is blistered because the solvent became entrapped in the coating while the temperature was very high; this causes _____. Since the entrapment of solvent prevents the coat from curing properly, another coating defect appears, which is called _____. When the temperature is hot, the coating can dry very quickly and the coating surface can skin over the underlayer; this is called _____ coating defect.

A. Crinkling, blistering, cratering, swelling

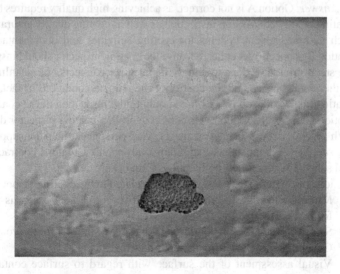

FIGURE 4.35 Coating defect. (Courtesy: Shutterstock.)

B. Wrinkling, blistering, solvent boil
C. Swelling, solvent boil, pinhole, wrinkling
D. Wrinkling, solvent boil, swelling, pinhole

Answer: Swelling is a kind of defect related to the presence of moisture on the substrate or applying wet-to-wet coating. The formation of blisters because of solvent entrapped in the coating is called solvent boil; if it is not removed, the solvent boil will turn into pinhole defect. The skinning of the coating over the underlayer in hot temperature is called wrinkling, so option C is the correct answer.

5. Which type of coating defect is illustrated in Figure 4.35?
A. Blistering
B. Swelling and peeling off
C. Pitting
D. Eye fish

Answer: The defect is very similar to blistering, but the better choice of term for this type of defect is swelling and peeling off. Cratering, also called pitting, is a kind of paint film defect in which small, rounded depressions are created in the coating. This type of defect is different from pitting, so Option C is not correct. Eye fish is not a correct answer, so option D is wrong. Therefore, option B is correct.

6. Which sentences are correct regarding coating inspection?
A. Coating inspection is the only element required to achieve coating quality.
B. The inspection of coating is limited to the coating itself.
C. Both surface cleanliness and roughness are important before coating application.
D. A coating inspector has various contacts in the projects.

Answer: Option A is not correct, as achieving high quality requires both coating quality control and inspection as well as coating quality assurance, such as management systems for coating activities and documentation. Option B is not correct either, because the coating inspector should inspect the steelwork that is done to remove metal surface defects, the cleanliness of the metal surface and the control of the climate condition. In fact, the coating inspector's job is not limited only to the inspection of the coating. Option C is correct. Option D is correct as well; a coating inspector deals with several parties, such as the owner; yard personnel; the paint supplier; the supplier of components like piping, valves and actuators; contractors and subcontractors. Thus, both options C and D are correct.

7. Which option indicates a task that is not related to the coating inspector?
A. Making sure that the mixing of the two-component epoxy is performed correctly
B. Selection and checking of the correct usage of thinner for the coating
C. Checking and measuring the coating film thickness
D. Visual assessment of the surface with regard to surface contaminants such as rust, oil, grease, etc.

Answer: Options A, C and D are all related to the coating inspector's task. However, option B is not completely correct. An inspector should make sure that the thinner is applied correctly during the painting application. However, selecting the coating thinner is not the job of the inspector normally. The manufacturer of the coating could propose the best thinner, or the thinner can be selected by a coating engineer who works for the client or contractor company.

8. Which sentences are correct with regard to coating inspection?
   A. Inspector level I has higher competency compared to Inspector levels II and III.
   B. A tape can be used to evaluate surface dust according to ISO 8502 part 3.
   C. A greater difference between the wet and dry thermometer temperature values in a sling psychrometer indicates higher humidity in the air.
   D. Some basic tools, like a mirror, magnifier and flashlight, can be used for surface and coating inspection.

   Answer: Option A is not correct because inspector level I has less competency compared to inspector levels II and III. Option B is correct because a tape is typically used according to ISO 8502 part 3 to assess the dust on a metal surface before coating application. Option C is not correct, because a greater the temperature difference between the wet and dry thermometer in a sling psychrometer indicates more dry weather and less humidity. In fact, more evaporation from the wet thermometer in dry weather results in a colder temperature registered by the wet thermometer and a greater temperature difference between the wet and dry thermometers. Option D is correct; a mirror, magnifier and flashlight are all considered basic visual inspection tools. Thus, options B and D are correct.

9. Which sentence is correct with regard to assessing the surface cleanliness of a substrate before coating application?
   A. Soluble salts are measured by a conductivity meter as per ISO 8502 part 9.
   B. A dew point calculator is used to assess condensate formation as per ISO 8502 part 5.
   C. Assessment of dust is done according to ISO 8502 part 4.
   D. All of the above sentences are wrong.

   Answer: Option A is correct; the measurement of soluble salts is performed according to ISO 8502 part 9. Option B is not correct, because a dew-point calculator is used to assess the possibility of condensate formation on a metal surface according to ISO 8502 part 4. Option C is not correct, since the assessment of dust is performed according to ISO 8502 part 3. Option D is not correct, because Option A is correct.

10. Fill in the gaps with the correct words.

    The most common method of measuring the DFT of coating on a metal surface is _____. Coating inspection is not limited to just coating; as an example, it is important to measure the amount of contaminants such as grease, oil and soluble salt from _____ according to the ASTM D4940 standard. Knife, pull-off and tape tests are used to evaluate the _____ of the coating. The other important test that is done to evaluate the paint _____ is a rub test with methyl ethyl ketone (MEK).

A. Eddy current, substrate, durability, curing
B. Electromagnetic, abrasives, adhesion, curing
C. Electromagnetic, substrate, durability, adhesion
D. Eddy current, abrasives, adhesion, adhesion

Answer: The most common coating DFT measurement method is electromagnetic. Measurement of abrasive contaminants such as grease, oil and soluble salt based on ASTM D4940 is used for sandblasting. Knife, pull-off and tape tests are used to evaluate the adhesion of the coating. A rub test with MEK is used to assess the curing of the coating. Thus, option B is the correct answer.

BIBLIOGRAPHY

1. American Society for Testing and Materials (ASTM) D3359 (2020). Standard test methods for rating adhesion by tape test. New York.
2. American Society for Testing and Materials (ASTM) D4752 (2020). Standard test method for measuring MEK resistance of ethyl silicate (inorganic) zinc-rich primers by solvent rub. New York.
3. American Society for Testing and Materials (ASTM) D4940 (2020). Standard test method for conductimetric analysis of water-soluble ionic contamination of blast cleaning abrasives. New York.
4. American Society for Testing and Materials (ASTM) D6677 (2020). Standard test method for evaluating adhesion by knife. New York.
5. Appleman, B., Drisko, R., & Neugebauer, J. (1997). *The Inspection of Coating and Linings, a Handbook of Basic Practice for Inspectors, Owners and Specifiers.* SSPC (The Society for Protective Coating): Pittsburgh, PA.
6. International Organization of Standardization (ISO) 11127 (2020). Preparation of steel substrates before application of paints and related products—Test methods for non-metallic blast cleaning abrasives—Part 5: Determination of moisture. Geneva, Switzerland.
7. International Organization of Standardization (ISO) 11127 (2011). Preparation of steel substrates before application of paints and related products—Test methods for non-metallic blast cleaning abrasives—Part 6: Determination of water-soluble contaminants by conductivity measurement. Geneva, Switzerland.
8. International Organization of Standardization (ISO) 8501 (2007). Preparation of steel substrates before application of paints and related products—Visual assessment of the surface cleanliness—Part 1: Rust grades and preparation grades of uncoated steel substrates and of steel substrates after overall removal of previous coating. Geneva, Switzerland.
9. International Organization of Standardization (ISO) 8501 (2007). Preparation of steel substrates before application of paints and related products—Visual assessment of surface cleanliness—Part 2: Preparation grades of previously coated steel substrates after localized removal of previous painting. Geneva, Switzerland.

10. International Organization of Standardization (ISO) 8501 (2006). Preparation of steel substrates before application of paints and related products—Visual assessment of surface cleanliness—Part 3: Preparation grades of welds, edges and other areas with surface imperfections. Geneva, Switzerland.
11. International Organization of Standardization (ISO) 8502 (2019). Preparation of steel substrates before application of paints and related products—Tests for the assessment of surface cleanliness—Part 3: Assessment of dust on steel surfaces prepared for painting (pressure-sensitive tape method). Geneva, Switzerland.
12. International Organization of Standardization (ISO) 8502 (2017). Preparation of steel substrates before application of paints and related products—Tests for the assessment of surface cleanliness—Part 4: Guidance of the estimation of the probability of condensation prior to paint application. Geneva, Switzerland.
13. International Organization of Standardization (ISO) 8502 (1998). Preparation of steel substrates before application of paints and related products—Tests for the assessment of surface cleanliness—Part 5: Measurement of chloride on steel surfaces prepared for painting (ion detection tube method). Geneva, Switzerland.
14. International Organization of Standardization (ISO) 8502 (2006). Preparation of steel substrates before application of paints and related products—Tests for the assessment of surface cleanliness—Part 6: Extraction of soluble contaminants for analysis—The Bresle method. Geneva, Switzerland.
15. International Organization of Standardization (ISO) 8502 (2020). Preparation of steel substrates before application of paints and related products—Tests for the assessment of surface cleanliness—Part 9: Field method for the conductometric determination of water-soluble salts. Geneva, Switzerland.
16. NORSOK M-501 (2012). *Surface Preparation and Protective Coating*, 6th edition. Lysaker, Norway.

5 Valve and Actuator Technology for the Offshore Industry

5.1 INTRODUCTION

Valves are essential components of piping systems and are used for different purposes, such as stopping or starting the fluid inside the piping, also called on/off application; regulating or controlling the fluid; preventing backflow; and performing safety functions. Valves can account for ~20% to 30% of the total piping cost. Industrial valves fail for various reasons, such as poor valve selection, lack of mechanical strength against pipeline loads or fluid pressure inside the piping, poor material selection, corrosion, inappropriate coating or wrong coating application, sealing failure, poor testing and inspection, etc. Failure of valve material and subsequent corrosion can occur both internally and externally. Internal corrosion and material failure of industrial valves are caused by contact with the internal fluid service. External corrosion of industrial valves related to the corrosive offshore environment could be due to wrong material selection or wrong coating selection and application. Failure of industrial valves because of corrosion has many negative consequences as mentioned in detail in Chapter 1.

It is essential to know that all valves, except for check valves and pressure safety valves, can be manual or automated. Manual valves are operated by an operator force that moves a lever or handwheel on the valve to open and close it (see Figure 5.1). However, some valves open and close automatically by means of actuators. Valve automation via actuator may be preferred for different reasons, such as ease of operation, fast operation or automatic operation. Some valves are located in hazardous areas where it is dangerous for personnel to go. Before discussing valves and actuators in more detail, it is helpful to be aware of the offshore field development options in order to have a better understanding of where offshore valves and actuators are located.

5.2 OFFSHORE FIELD DEVELOPMENT OPTIONS

This section discusses different offshore oil field development options, including both topside and subsea.

DOI: 10.1201/9781003255918-5

FIGURE 5.1   Manual valve operated by an operator. (Courtesy: Shutterstock.)

FIGURE 5.2   Fixed platform in the Mediterranean Sea close to Italy. (Courtesy: Shutterstock.)

5.2.1 TOPSIDE

Topside offshore refers to units that are located on platforms or ships in seas and oceans. Fixed platforms are often used for the development of offshore fields in shallow water depths. They could be made of steel or concrete and rely on their own weight to sit on the seabed or ocean floor. An oil platform is a large structure used to house workers, facilities and units, including industrial piping, valves and actuators, that are used for the production and treatment of oil, gas and other petroleum byproducts. Figure 5.2 illustrates a fixed platform. Fixed platforms can be used for a design life of +30 years, and a maximum of 300 m water depth, as illustrated in Figure 5.3.

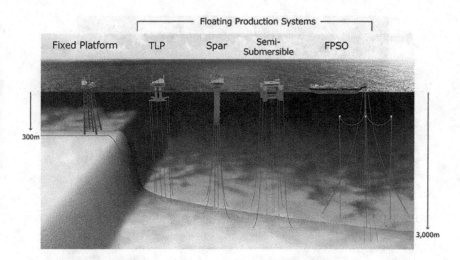

FIGURE 5.3 Oilfield development options. (Courtesy: MODEC.)

The alternative to fixed platforms are floating platforms, which can be used for developing oil and gas fields in deeper depths. Again, all the facilities, including piping, industrial valves and actuators, are located atop the floating facilities. Tension-leg platforms (TLPs), spar and semi-submergible platforms are all known as floating platforms. The use of ships, called floating production storage and offloading (FPSO) units, is becoming more common and popular for oilfield development. Figure 5.3 compares all the different topside options for oilfield development, including fixed and floating platforms as well as FPSOs. TLPs and extended tension-leg platforms (ELTPs) are typically used for water depths of more than 300 m to 1.5 km. TLPs are constructed with columns and pontoons beneath the platform itself. Pontoons are horizontal steel structures that connect the columns; they are installed at the bottom of the platform. TLPs are permanently moored or secured to the seabed or ocean floor by tendons or tension legs that keep the platform in place and support the floating platform structure. They have a buoyant hull that supports the topside platform. TLPs have been used in the oil and gas industry since the 1980s for both drilling and production purposes. Figure 5.4 illustrates a TLP platform at sea during sunset.

A spar is another type of floating platform typically used in deeper waters than those in which TPLs are used. Unlike TLPs, which have a four-column design, a spar platform consists of a large diameter, single vertical cylinder that is designed to support the deck (see Figure 5.5). Like TLPs, spar platforms are moored or fixed to the sea or ocean bed. However, the cylinder of the spar platform does not extend all the way to the seafloor. This large cylinder instead serves to stabilize the platform in the water, allowing movement of the spar platform, and absorbing forces and loads applied to the platform, such as hurricane winds and waves. Again, in this case, all the facilities, including the piping, valves, actuators and process units, are located on top of the spar platform.

FIGURE 5.4   TLP platform (Courtesy: Shutterstock.)

FIGURE 5.5   Spar platform in the Gulf of Mexico. (Courtesy: Shutterstock.)

The fourth type of platform is semi-submersible, which is a special type of platform with legs or pontoons that provide buoyancy for the platform to float, and weight to keep the structure upright. Columns in the platform are used to provide stability, and pontoons connect the columns together. This type of platform, illustrated in Figure 5.6, can be used for multiple purposes, such as offshore drilling rigs and oil production. It is noticeable that the old semi-submersible platforms could have multiple purposes, but the newly designed, semi-submersible platforms are specifically designed for a unique purpose. As illustrated in the figure, this type of platform has large columns that are connected to large pontoons. A semi-submersible platform has some advantages, such as providing a large deck, exceptional stability on the sea and ocean, the possibility of being moved from one location to another, etc. This type of platform can be used for oil field development as deep as almost 3,000 m. All facilities, including valves, actuators and piping systems, are located topside on the top or deck of the jacket, as illustrated . in Figure 5.6.

FIGURE 5.6   Semi-submersible platform. (Courtesy: Shutterstock.)

FIGURE 5.7   FPSO. (Courtesy: Shutterstock.)

An FPSO is a kind of ship that can produce, store and offload produced hydro-carbon, such as oil, to a pipeline or tanker. Oil and gas go through a series of processes, which are all done on board the ship, before being transferred to storage. As illustrated in Figure 5.7, an FPSO is used to develop and produce oil and gas in subsea areas at depths of 3,000 m or more. The first step is to drill the subsea well and transfer the three phases of oil, gas and water from the wellhead to the FPSO through flowlines. The first step of production on the platform or FPSO takes place in a separation unit. A separator is a type of pressure vessel used to separate the three phases of oil, gas and water from each other. The separation can be achieved in one, two or three stages, in which one, two or three separators are used for the separation process, respectively. The produced oil can be stored for a period of time before being transported from the separator to a pipeline or tanker. If the amount of produced oil is high, then a pipeline is a more suitable choice for further oil transportation from the FPSO. All the piping, valves and actuators in this case are located topside, on the deck of the FPSO. Since some valves and actuators are used subsea and thus immersed in the water, the next section provides some general information about subsea oil and gas development.

5.2.2 SUBSEA

Subsea engineering is comprised of different activities, such as exploration, drilling, completion and production under the sea. Subsea engineering and development are required to design the components, facilities and structures that will be used underwater on the seabed. The concept of subsea engineering and oilfield development began in the early 1970s with the installation of wellheads, Christmas trees, production components and facilities such as manifolds, jumpers and pipeline on the seabed. Figure 5.8 illustrates a typical subsea production system.

A subsea well is drilled down through the seafloor to reach the oil. After drilling the well in the seabed and placing tubing inside it for oil production, the well should be completed. Well completion is the process of making the well ready for production after the drilling has been completed. As part of this process, casings should be installed inside the well to prevent it from collapsing. The area between the casing and tubing is sealed by a tubing packer. Packers are used to prevent the pressure of the well fluid from entering the production casing. A wellhead is installed on top of the well on the seabed; it serves multiple purposes, such as providing a pressure barrier, providing access to the well bore and production tubing, ensuring pressure and structural integrity for the well casing and tubing, etc. The upper part of the wellhead, which is located on the subsea bed, contains valves, piping, actuators, fittings and the structures necessary to control the flow and pressure inside the well; this is called a "Christmas tree" or simply a "tree." It is called a wet tree (see Figure 5.9) if it is installed subsea and a dry tree if it is installed on land. In addition to the valves and actuators that are installed on

FIGURE 5.8   Typical subsea production system. (Photograph by author.)

FIGURE 5.9   Subsea tree. (Courtesy: Shutterstock.)

the trees, some valves are installed on the subsea manifolds and in the well. A subsurface safety valve is installed in the well on the upper wellbore to provide emergency stoppage of well production in the event of a high-pressure scenario inside the well. This type of valve is excluded from the scope of this book.

Subsea umbilicals, also called umbilical cables or umbilical power, are connected between the subsea structures and surface platform, as illustrated in Figure 5.10. An umbilical, as illustrated in Figure 5.10, is a combination of hoses and cables used to transfer electricity and hydraulic power for subsea operation; some subsea components and facilities such as actuators work with hydraulic force and some work with electricity. Therefore, it is important to transmit both hydraulic force and electricity from the topside platform to the subsea equipment.

The other important component is a jumper. A jumper is a small piece of pipe that connects two structures together. For example, a jumper may connect a wellhead to a manifold, two manifolds together or a manifold to a pipeline end terminal (PLET). A manifold is the combination of a structure, piping and valves that are used to simplify the subsea system and reduce the number of subsea pipelines and flowlines. Manifolds, in fact, are designed to combine, distribute and control the flow of the fluid produced from the well. Subsea manifolds (see Figure 5.11) can be connected to different wells; they gather all the fluid from various wells and convert all of the inline piping into one or two outlet pipes. A prefix such as "two slot," "four slot," "six slot" or "eight slot" may be used for the manifold. The number of slots represents the number of wells that are connected to the manifold. As an example, a four-slot subsea manifold is connected to four subsea wells and trees. Using manifolds can reduce the amount of piping and the number of pipelines needed subsea. Piping and valves inside subsea manifolds are responsible for controlling and directing the flow of the produced fluid.

A flowline is a subsea pipeline that transports the fluid from a tree or manifold to the surface facilities and platform. It is important to know that a flowline is a

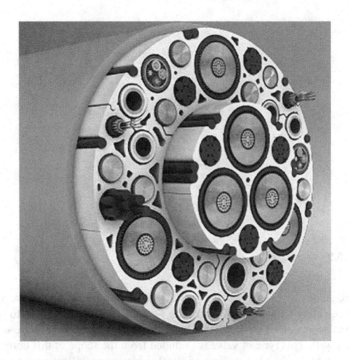

FIGURE 5.10   Subsea umbilical. (Photograph by author.)

FIGURE 5.11   Subsea manifold. (Courtesy: Shutterstock.)

FIGURE 5.12   Subsea flowline on the seabed approaching a platform. (Courtesy: Shutterstock.)

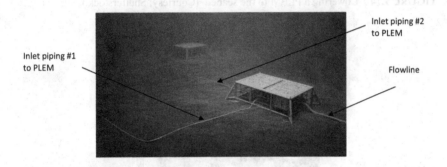

FIGURE 5.13   PLEM with two inlet pipes and a flowline outlet. (Courtesy: Shutterstock.)

kind of flexible pipeline that moves on the seabed (see Figure 5.12). In order to connect a flowline to the top of a platform, another pipe called a *riser* should be connected between the platform and the flowline.

In addition to combining, distributing and controlling the fluid, manifolds can provide structural support for pipe-to-pipe connections, such as a PLET. Termination units, which are important parts of subsea field development, include both a PLET and a pipeline end manifold (PLEM). A PLET is a type of manifold or structure that connects a single pipe to one another pipe. A PLET can be used when a riser is connected to a flowline. A PLEM is a simple manifold structure that is used to connect a rigid pipeline to another structure or pipe. The next section provides more information about the valve and actuator technology used in the offshore industry. Figure 5.13 illustrates a manifold that acts as a PLEM and contains two inlet pipes and one outlet flowline. The inlet piping can come from manifold(s) and/or tree(s). In fact, the manifold acts as a structure for the termination of the flowline (pipeline).

FIGURE 5.14 Lowering a PLEM to the seabed. (Courtesy: Shutterstock.)

FIGURE 5.15 SDU. (Courtesy: Shutterstock.)

Figure 5.14 illustrates a PLEM being lowered into the seawater and down to the seabed.

The other type of manifold is a subsea distribution unit (SDU); it contains many small valves on the piping, which are used to transport chemicals to the production manifolds. The types of chemicals that are distributed through an SDU are wax and scale inhibitor, hydraulic fluid, mono-ethylene glycol (MEG) and corrosion inhibitor. Figure 5.15 illustrates an SDU from the outside; the piping and valves are located inside the SDU.

Wax can be deposited from the oil inside the pipe and cause operational problems, such as pressure drop and reduction in flow rate; to prevent this, wax inhibitor is injected into the piping. Scale forms inside the piping from the produced water, especially when the water contains minerals that can build up inside the piping and cause pressure drop and flow assurance problems. Hydraulic fluid is mainly supplied to hydraulic actuators. MEG is a chemical injected into gas service to prevent hydrate formation inside the piping system. MEG absorbs water and moisture from the gas in a process called gas dehydration. It is not desirable to transport gas with water, as water in low temperatures and under high pressure can make ice-like crystalline compounds called gas hydrates or simply hydrates. Hydrates can build up in the pipe and block the pipe completely.

5.3 INDUSTRIAL VALVES FOR THE OFFSHORE INDUSTRY

5.3.1 VALVES FOR TOPSIDE

5.3.1.1 On/Off Valves

On/off valves are used to start and stop the fluid; these types of valves are divided into four main categories in the offshore sector: ball valves, gate valves, plug valves and butterfly valves. Gate valves are divided into two categories: through conduit gate (TCG) valves and wedge-type gate valves. The four main types of valves provide single isolation. The other type of valve, which is used for double isolation, is called a modular valve. The next section explains ball valves in more detail.

5.3.1.1.1 Ball Valves

Ball valves have a metal ball with a hole inside it; the ball is placed inside the valve body between two seats. The ball, which is called a closure member, moves 90° between open and closed positions. Ball valves are not suitable for *fluid control*, also called *throttling*, which would necessitate keeping the ball in an intermediate position between open and closed. The reason why a ball valve cannot be used for this purpose is that the valve would be subject to excessive wear and damage. Figure 5.16 illustrates a ball valve in open position; here, the hole inside the ball is aligned with the flow direction, so the fluid can pass through the valve. The half ring in the picture, between the valve shell or body and the ball is called the seat. The other important component of the valve is the stem, which is connected to the valve ball from one side and to the valve operator (gear or actuator) from the other side. The function of the stem is to transfer the load from the operator to the valve closure member in order to open and close the valve. Ball valves are known as *quarter-turn* valves, in which the ball and stem move only a quarter turn or 90° between open and closed positions. The quarter-turn movement of the ball is an advantage since it provides a fast operation feature and capability to a ball valve.

The body or shell of valves, including ball valves, is the main pressure-containing part that provides sealing between the fluid and the external environment. Any failure of the body leads to the emission of internal fluid to the environment. In

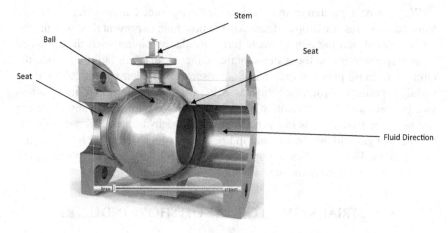

FIGURE 5.16   Ball valve in open position. (Photograph by author.)

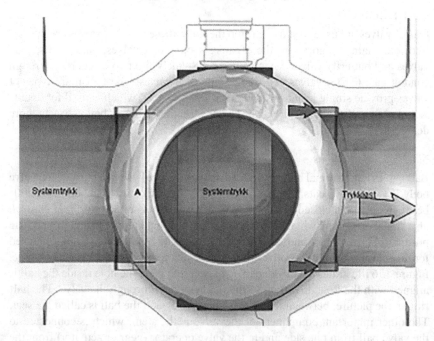

FIGURE 5.17   Ball valve in closed position. (Photograph by author.)

addition to the body, the stem and the bolting (bolts and nuts) are also considered pressure-containing parts. Bolts in ball valves are used to connect the body pieces or body and bonnet together. A bonnet is a valve component that is installed on the body as a cover. Figure 5.17 illustrates a ball valve in a closed position, in which the ball hole is perpendicular to the flow direction. Coating is mainly applied on the body and bonnet of ball valves.

Ball valves can be used in both clean and dirty or particle-containing fluids. The main difference between a ball valve used in clean versus particle-containing fluids is related to the seat material and design; the seat of a ball valve in dirty services is metallic, while a soft seat is used in clean services. A metal-seat ball valve may be selected for other reasons, such as applications prone to high temperatures and significant pressure drop. A soft or non-metallic seat could be damaged in dirty fluid services, and in applications in which pressure drop and high temperatures are prevalent. A soft-seat ball valve could be selected as the first choice in other applications, since it is cheaper than a metal seat, unless at least one of the above-mentioned conditions with regard to fluid cleanliness, high temperature and pressure drop occurs. Figure 5.18 illustrates a metal-seat ball valve in a half-open position. Figure 5.19 illustrates the disassembly of a small soft-seat ball valve used for water service. Two white seats, made of Teflon, are located on the sides of the ball.

FIGURE 5.18   Metal-seat ball valve. (Courtesy: Shutterstock.)

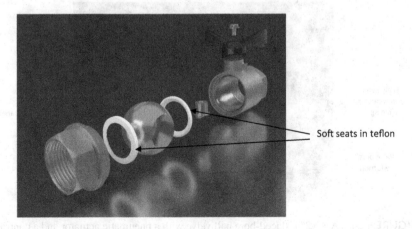

FIGURE 5.19   Soft-seat ball valve disassembly. (Courtesy: Shutterstock.)

Ball valves can be full-bore or reduced-bore. Figure 5.20 illustrates a full-bore ball valve with a straight flow path inside the valve; the valve's internal bore is the same as both ends of the valve, which are connected to piping. A full-bore valve is also called full port, alternatively. The main advantage of a full-bore ball valve is minimum pressure drop and fluid capacity loss across the valve and less wearing of the valve internals.

In a reduced-bore ball valve, the valve internals and inside diameter of the valve are smaller than the connected piping, as illustrated in Figure 5.21. Figure 5.21 shows a reduced-bore ball valve with 3″ flanges that are connected to piping and a reduced bore of 2″. The reduction of the valve bore can be distinguished

Bore or port of the valve is full where fluid flows without reduction

FIGURE 5.20   Full-bore ball valve. (Photograph by author.)

Actuator

Flange end connected to piping

Control panel

Valve bore reduction

FIGURE 5.21   A 3″x2″ reduced-bore ball valve with a pneumatic actuator and a control panel. (Photograph by author.)

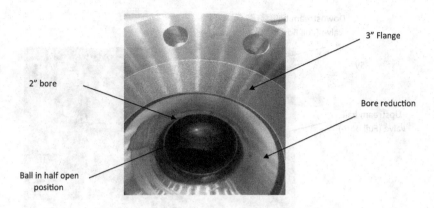

3" Flange

2" bore

Bore reduction

Ball in half open
position

FIGURE 5.22   Side view of a 3"x2" ball valve. (Photograph by author.)

by looking at the valve from the outside, as shown in the figure. The illustrated valve is automated with a pneumatic (air) actuator. A control panel is attached to the actuator to command, control and direct the air or pneumatic force from the actuator to the valve for valve operation. Figure 5.22 provides a side view of the same valve. The ball shown in the figure is 2" and is in almost half-open position. A reduced-bore ball valve is designated and identified with two sizes with an "x" between. The first size is the larger size or the size at the point of connection to the piping, and the second size is the smaller size, the size of the valve internal (e.g., 3"x2").

A reduced-bore ball valve has a smaller ball and smaller internals and requires less force or torque for opening and closing compared to a full-bore ball valve. A reduced-bore ball valve is typically considered the first choice for ball valve selection since it is more economical and cheaper than a full-bore ball valve. Valves are normally selected by valve and process engineers based on different parameters, like size, pressure, temperature, fluid type, application, etc. Although reduced bore is preferred over full bore for economic reasons, some conditions do not allow the usage of reduced-bore ball valves. The first condition is pressure drop and wearing inside the valve. If the pressure drop produced by a reduced-bore valve is high enough to cause flow assurance problems, then a reduced-bore ball valve should not be used. Process engineers are the correct references to give feedback on the acceptance of reduced-bore ball valves with regard to pressure drop.

The second case in which full-bore ball valves are required to prevent pressure drop is when the ball valves are used before and after a pressure safety valve (PSV). PSVs are installed on pressure piping and equipment to release the overpressure fluid to the flare lines. In fact, PSVs are known as the final safety solution to release overpressure accumulation in pressurized equipment and prevent the pressurized system from failing when other safety systems fail to function. The function of PSVs is explained in more detail later in this chapter. Typically, more than one PSV is designed and selected for a system; PSVs are subject to

Downstream ball
valve (Full Bore)

PSV

Upstream ball
valve (Full Bore)

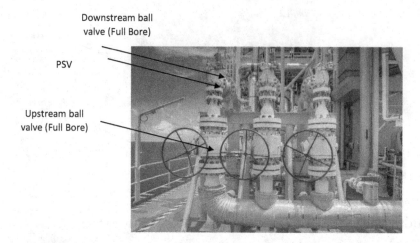

FIGURE 5.23 PSVs and full-bore ball valves on a compressor discharge line. (Courtesy: Shutterstock.)

maintenance and calibration, so when one PSV is removed from the piping system, the other can be used as a backup. Each PSV is located between two ball valves—one upstream (before) and one downstream (after). The ball valves installed before and after a PSV in operation are always open. However, when a PSV is removed from the piping system for maintenance or modification, the ball valves before and after it should be closed. In this case, the backup PSV is used, and the ball valves located before and after the backup PSV should be moved from closed to open position. All ball valves before and after PSVs should be full-bore to facilitate the release of overpressure gas or liquid to the flare system. Figure 5.23 illustrates three parallel PSVs; the ball valves located before and after them are installed on a compressor discharge line.

The third case in which the valve bore should be completely the same as the pipe bore is due to piping-injected gadget (PIG) running. PIG running is common in pipelines, flowlines and subsea manifold headers. A PIG (see Figure 5.24) is a tool that is sent through the pipeline and moved forward by the pressure of the fluid in the pipeline. A PIG may be sent inside a pipeline for various purposes such as cleaning and inspection. In case of PIG running, a pipeline valve should have exactly the same internal diameter as the connected pipe to facilitate the passage of the PIG without any obstruction. In this case, the ball valve is not considered either full or reduced bore, but instead a ball valve with a special bore.

The other important ball valve feature is the choice between a floating or trunnion ball valve. A floating ball valve contains an unsupported ball, while a trunnion ball valve has a trunnion or support under the ball. The trunnion support could be a plate or flange. As a general rule, floating ball valves are common for smaller and low-pressure class valves that require less force or torque for operation. The ball valves illustrated in Figures 5.16 and 5.17 are floating ball.

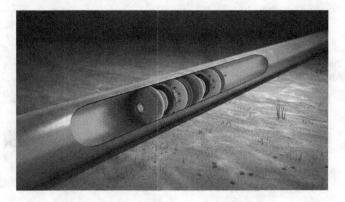

FIGURE 5.24   PIG in a subsea flowline. (Courtesy: Shutterstock.)

FIGURE 5.25   Trunnion-mounted ball valve disassembly. (Photograph by author.)

Figure 5.25 illustrates the disassembly of a ball valve when the ball and its support trunnion have been removed from the valve body.

The other important design feature for ball valves is side- or top-entry design. Side-entry design is more common for topside ball valves. In this model, access to the valve internals is made possible from the side by unscrewing the bolts between the body pieces. Figure 5.26 illustrates a manual ball valve that is operated with a handwheel and gearbox located on an offshore platform. The ball valve is in carbon steel material coated with zinc-rich primer and zinc-epoxy coating. The ball valve in the figure is side entry, as assembly or disassembly of the valve is performed from the side by unscrewing the side flange bolts.

Side body of
the valve
(side entry)

FIGURE 5.26   Side-entry carbon steel ball valve in offshore coated with zinc-rich coating. (Courtesy: Shutterstock.)

   Top-entry ball valves have a bonnet or cover that is typically bolted to the body of the valve. Access to the valve internals is possible from the top by unfastening the bonnet and body bolts. Top-entry ball valves have one main advantage: the ability to perform online maintenance means that there is no need to disassemble the valve from the connected piping for maintenance, in contrast to the side-entry design. The bonnet is a heavy component that adds significant weight to the valve so, in general, top-entry ball valves are heavier than side-entry due to their bulkier bonnet. Top-entry valves have more advantages compared to side-entry design, such as more resistance against the applied loads on the valve, such as loads from connected piping or pipeline systems, and more flexibility to enlarge the valve stem in case higher strength is required and expected from the valve stem. Top-entry ball valves used topside are typically welded to the connected piping, unlike side-entry design, which is typically flange-end connection. Welding top-entry ball valves to the pipeline saves using flange connections and reduces costs. Figure 5.27 illustrates a large, heavy top-entry ball valve installed on a 38″ pipeline. The ball valve body material is carbon steel, and the operating temperature is around 70°C. The valve is coated with a white, zinc-rich coating system according to Norsok M-501 coating system 1.

5.3.1.1.2   Gate Valves

5.3.1.1.2.1 Through Conduit Gate Valves   This section explains in more detail the different types of gate valves that are used in the topside oil and gas industry. TCG valves are used for on/off or to start/stop the fluid. Like ball valves, TCG valves are not proposed for fluid control or intermediate positions, which can lead to severe erosion and wearing inside the valve. TCG valves are better choices compared to ball valves in sandy or particle-containing services. However, ball valves do have some advantages in comparison with gate valves. Ball valves are easier and faster to open and close compared to gate valves, with just a 90° rotation of the ball. Gate valves could have greater height compared to ball valves, especially in large sizes of 12″ and above. This extra height could be a problem

FIGURE 5.27 A 38″ top-entry ball valve for a pipeline on a topside platform. (Photograph by author.)

when there is a vertical space limitation. In addition, gaining access to the top of a large gate valve for operation could be difficult, so a platform may be required to provide the access needed for valve operation. The other disadvantage of a gate valve compared to a ball valve is related to the friction between the stem and its sealing which is higher in a gate valve compared to a ball valve due to the linear movement of the stem in a gate valve. Higher friction between the stem and stem sealing in a gate valve causes wearing and damage to the stem seals and damage to the stem eventually. TCG valves could be full-bore or reduced-bore, like ball valves. Full-bore TCG valves are the preferred choice, as these valves are used in dirty services. Having a full bore facilitates the passage of particles and dirt inside the valve. Unlike ball valves with a rotary quarter-turn or 90-degree ball and stem rotation, gate valves have a linear stem and closure member (disk) motion which moves up and down. All types of gate valves have two seats, like standard ball valves. The seats of gate valves are metallic as a standard design. TCG valves have a top-entry design, meaning that they have a bonnet or cover on top of the body. The body and bonnet of gate valves are bolted together in most cases. The valves' internals may be accessed from the top for maintenance or other purposes by unscrewing the bolts fastening the body and the bonnet together. TCG valves are divided into slab gate and expanding gate valves. Figure 5.28 illustrates a manual slab gate valve in an open position, as the hole inside the disk is aligned with the fluid passage inside the valve. If the operator turns the handwheel clockwise, the stem will move down and push the disk downward and the valve will close. Figure 5.29 shows the stem, seats and disk of a slab gate valve in a closed position. As shown in the figure, the fluid direction is from left to right, and the sealing between the disk and downstream seat is achieved by means of fluid pressure. Figure 5.30 illustrates a disassembled slab gate valve with key parts highlighted, such as the body, bonnet, stem and disk.

Double-expanding gate valves are more expensive than slab gate valves. Instead of having a one-piece disk, double-expanding gate valves contain two

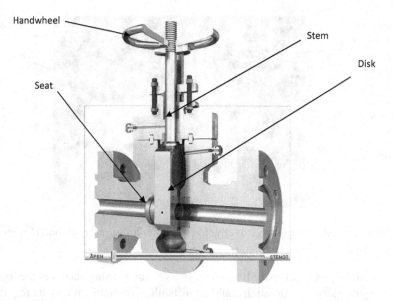

FIGURE 5.28   Manual slab gate valve in open position. (Photograph by author.)

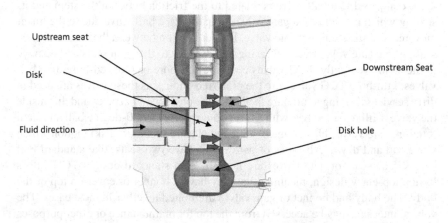

FIGURE 5.29   Manual slab gate valve in closed position. (Photograph by author.)

half disks—one male and one female. Figure 5.31 illustrates a double-expanding gate valve in an open position. The handwheel is moved clockwise to move the stem and disk downward and close the valve. In both open and closed positions, the linear force of the stem is transmitted to the half disks, and both half disks are pushed against both seats to provide sealing for the valve (see Figure 5.32).

The lateral expansion of the half disks due to stem force is called the wedging effect. The other type of expanding gate valve is a single-expanding gate valve, in which the two half disks only expand due to the wedging effect in the closed position. Single-expanding valves can be used instead of double expanding in order

FIGURE 5.30   Disassembled slab gate valve. (Photograph by author.)

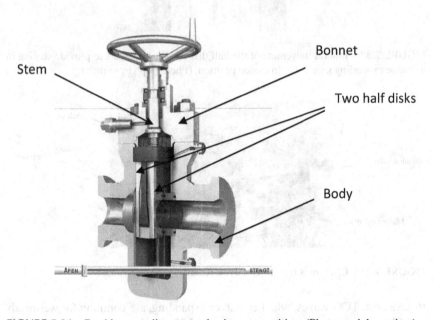

FIGURE 5.31   Double expanding gate valve in open position. (Photograph by author.)

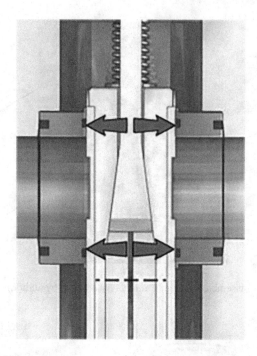

FIGURE 5.32   Lateral movement of the half disks toward the seats to provide sealing in a double expanding gate valve in closed position. (Photograph by author.)

FIGURE 5.33   Christmas tree TCG valves. (Courtesy: Shutterstock.)

to save cost. TCG valves, whether slab or expanding, are common for wellheads and Christmas trees, as illustrated in Figure 5.33. All the valves shown in the figure on the Christmas tree are TCG valves. The produced fluid from the reservoir passing through the Christmas tree contains a lot of sand; TCG valves are the best choice for such a dirty service. The lower valve, which is not completely shown in the figure, is called a master valve. According to API Spec. 6A, "Wellhead and tree equipment," identical to ISO 10423, two master valves are typically located at the bottom of the wellhead. The lower master valve, as shown in the API 6A is manual, is normally open, allowing the production fluid to pass through the tree. The upper master valve is called a safety valve in API 6A; it is an actuated valve that closes the production to protect the downstream equipment if something goes wrong in the well. The valve on the right-hand side is called a wing valve, which can be manual or actuated. A wing valve is another valve that can stop the flow production and isolate the downstream facilities and piping. The valve on the left-hand side, which is manually operated, is called a kill wing valve. It is used to inject chemicals into the well. A swab valve is the one on the top, providing vertical access to the well, typically for well intervention. Well intervention, or well work during the well design life, may include maintenance, cleaning or placing equipment such as a pump inside the well to increase production.

5.3.1.1.2.2 Wedge Gate Valves Wedge gate valves, like other types of gate valves (slab, expanding), are used for on/off purposes and for flow isolation. The gate of a wedge-type gate valve is in the shape of a wedge. A wedge gate valve, again like other types of gate valve, is not recommended for throttling or fluid control, as it can create excessive wearing and erosion on the valve internals such as the disk, seats and lower part of the stem. In addition, keeping a wedge gate valve in a semi-open position leads to chattering of the disk. A wedge gate valve, like an expanding gate valve, is a torque-seated valve, meaning that the sealing between the valve disk and seats is achieved through stem force and not fluid pressure. The disadvantages mentioned earlier in this chapter for gate valves in comparison to ball valves with regard to height, slower speed of operation and more stem sealing wear are all true for wedge gate valves. Wedge gate valves have other disadvantages compared to TCG valves; they are not suitable for particle-containing services, as sand and particles can accumulate at the bottom of the wedge and damage it during closing and/or prevent the valve from closing completely. Figure 5.34 illustrates a solid-wedge gate valve in the closed position on the right and the wedge of the valve on the left. The handwheel should be

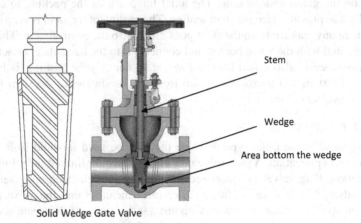

Stem

Wedge

Area bottom the wedge

Solid Wedge Gate Valve

FIGURE 5.34 A solid-wedge gate valve. (Courtesy: Hardhat Engineering.)

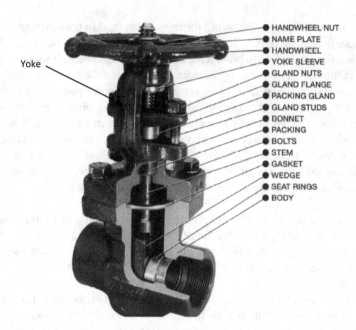

● HANDWHEEL NUT
● NAME PLATE
● HANDWHEEL
● YOKE SLEEVE
● GLAND NUTS
● GLAND FLANGE
● PACKING GLAND
● GLAND STUDS
● BONNET
● PACKING
● BOLTS
● STEM
● GASKET
● WEDGE
● SEAT RINGS
● BODY

Yoke

FIGURE 5.35   Wedge gate valve with parts list. (Photograph by author.)

rotated counter-clockwise to lift the stem and wedge so the valve opens and the fluid passes through the valve.

Figure 5.35 illustrates a wedge gate valve and part list. Bolts are used to connect the body and bonnet. A gasket is a type of sealing placed between the body and bonnet to prevent leakage. Packing is the stem sealing, which is typically in graphite. There is a gland located on the top of the packing and below the gland flange. The gland flange is tightened with bolts and nuts, which provide an axial force on the gland and packing. The axial force causes the packing to expand radially and provide effective stem sealing. The sealing of the stem is very important since any leakage from the stem goes directly to the environment. The yoke is integrated with the valve bonnet and connects it to the handwheel or actuator. The yoke sleeve, also called the yoke nut, stem nut or yoke bushing, is located around the stem and transfers the load from the handwheel to the stem for the linear movement of the stem.

5.3.1.1.3 Plug Valves

A plug valve is the other type of valve that can be used for fluid isolation or fluid on/off application. A plug valve may also be used for limited throttling fluid application. Plug valves are proposed for dirty or particle-containing services. Plug valves, like ball valves, are quarter turn, meaning that the valve moves between open and closed positions with just a 90° rotation of the steam and plug (closure member). The plug (item H in Figure 5.36) has a cylindrical or conical tapered shape with a hole inside. When the hole of the plug is in line with the

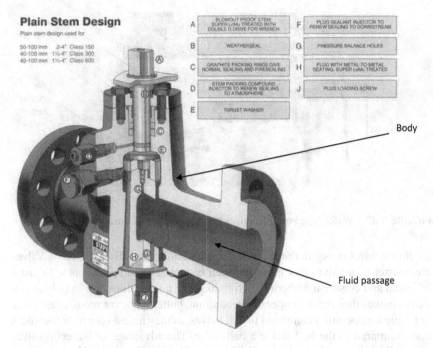

Plain Stem Design

Plain stem design used for

50-100 mm 2-4" Class 150
40-100 mm 1½-4" Class 300
40-100 mm 1½-4" Class 600

A	BLOWOUT PROOF STEM. SUPER LoMu TREATED WITH DOUBLE D DRIVE FOR WRENCH
B	WEATHERSEAL
C	GRAPHITE PACKING RINGS GIVE NORMAL SEALING AND FIRESEALING
D	STEM PACKING COMPOUND INJECTOR TO RENEW SEALING TO ATMOSPHERE
E	THRUST WASHER
F	PLUG SEALANT INJECTOR TO RENEW SEALING TO DOWNSTREAM
G	PRESSURE BALANCE HOLES
H	PLUG WITH METAL-TO-METAL SEATING, SUPER LoMu TREATED
J	PLUG LOADING SCREW

Body

Fluid passage

FIGURE 5.36 Lubricated plug valve in open position including part list. (Courtesy: MMK Engineering.)

flow (see Figure 5.36), the valve is open and the flow passes through the valve. When the plug is rotated 90°, the hole inside the plug is perpendicular to the flow direction and the solid part of the plug is placed in front of the flow, so the valve closes. Plug valves have metal-to-metal contact between the plug and body, but sometimes lubricant can be injected onto the plug for better sealing capability and to reduce metal-to-metal friction. Plug valves with lubrication are called lubricated plug valves. Plug valves, like TCG valves, are a suitable choice for particle-containing fluid. The advantage of plug valves over TCG valves is that, generally speaking, plug valves are more compact. However, plug valves can rely on lubricants for sealing, which is not desirable from an operation point of view. In fact, an operator should inject the lubricator into the plug valve in a specific period. Non-lubricant plug valve designs do exist; as an example, Teflon-sleeved plug valves have a Teflon sleeve that surrounds the plug and fits between the plug and body for sealing. However, Teflon is a soft (non-metallic) material and is not suitable for particle-containing services. Therefore, Teflon- or PTFE-sleeve plug valves are not suitable for dirty services because of the possibility of damage to the PTFE sleeve.

5.3.1.1.4 Butterfly Valves

Butterfly valves are the other group of valves that are used for flow isolation or on/off application. But it is important to know that butterfly valves can be used

Pipe

Flange

Wafer butterfly valve

Lever

Flange

FIGURE 5.37 Wafer-type butterfly valves between flanges. (Courtesy: Shutterstock.)

for flow control or regulation as well. Like ball and plug valves, butterfly valves are quarter-turn valves that move between open and closed positions with just a 90° stem and closure member (disk) rotation. The quarter-turn design of butterfly valves makes them easy to operate. In addition, butterfly valves require less force or torque for operation compared to ball valves, as they have a relatively low-mass disk compared to the ball inside a ball valve. The advantage of butterfly valves over ball, gate and plug valves is that they are lighter, cheaper and consume less space. In addition, butterfly valves can be designed and produced in a wafer-type (flangeless body) design, which reduces the weight of the valve significantly. Figure 5.37 illustrates wafer-type butterfly valves, shown in blue, placed between two flanges in a piping system. All the valves in Figure 5.37 are in open position since the levers are parallel to the pipe. Figure 5.38 compares a wafer-type rubber-lined butterfly valve in open and closed positions. Figure 5.39 shows a butterfly valve with flange connections on both sides, which is heavier and takes up more space compared to a wafer-type (flangeless body) butterfly valve.

The maintenance cost of butterfly valves is lower compared to the other types of valves since butterfly valves have a simple design and fewer moving parts. Butterfly valves are always reduced bore, so they create more pressure drop in the piping systems compared to ball and gate valves. Butterfly valves are not as robust as ball and TCG valves, so the usage of butterfly valves offshore could be limited to utility services such as air and water in low-pressure classes. Low-pressure classes are defined according to ASME B16.34, standard for valves, as CL150 and CL300, which are equal to 20 and 50 bar, respectively. The working principle of butterfly valves is simple; the load is transferred to the valve stem from the handwheel, lever or actuator, and rotates the stem. The stem is connected to the disk inside the valve, so the rotation of the stem is transferred to the disk. The disk and stem rotate together from open to closed position or vice versa. As mentioned above, butterfly valves can be used for flow control or regulation, in which case the disk is placed in an intermediate position between open and closed.

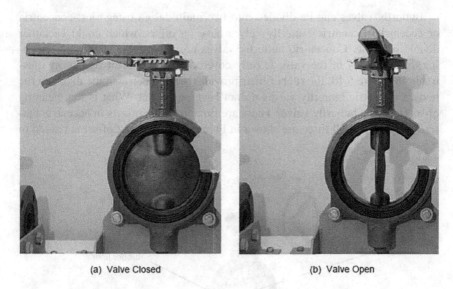

(a) Valve Closed   (b) Valve Open

FIGURE 5.38 Rubber-lined wafer-type butterfly valve in open and closed positions. (Photograph by author.)

FIGURE 5.39 Flanged connection butterfly valve. (Courtesy: Shutterstock.)

Butterfly valves can be divided into two main design categories: concentric or eccentric. Eccentric butterfly valves have an offset, which could be either double or triple. Concentric butterfly valves are typically rubber-lined, meaning that the body of the valve is fully covered with a layer of rubber, as illustrated in Figure 5.40; the rubber liner provides sealing between the disk and the body. Concentric butterfly valves do not have any offset. What is the meaning of "offset" in a butterfly valve? There are three possible offsets in eccentric butterfly valves, and all three are shown in Figure 5.41. The first offset is related to

FIGURE 5.40   Rubber-lined (concentric) butterfly valve. (Courtesy: Shutterstock.)

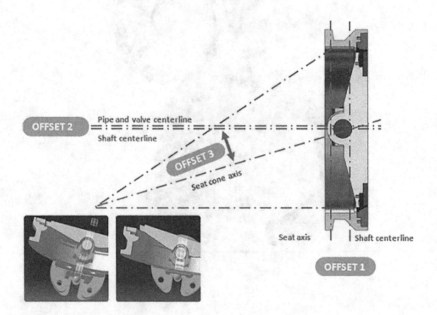

FIGURE 5.41   Butterfly valve offsets. (Photograph by author.)

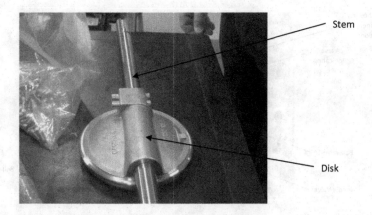

FIGURE 5.42 Stem-to-disk offset. (Photograph by author.)

the position of the disk in relation to the body. The disk of a concentric butterfly valve is in the middle of the body, whereas the disk of an eccentric butterfly valve is off to one side. The second offset has to do with the position of the stem to the disk. The stem passes exactly through the middle of the disk in a concentric butterfly valve. However, the stem does not pass through the middle of the disk (see Figure 5.42) in double or triple-offset butterfly valves. The third and last offset is related to the offset of the seat to the disk during sealing. A double-offset butterfly valve has the first and second offsets, whereas a triple-offset butterfly valve has all three offsets. Double- and triple-offset butterfly valves in general are more robust compared to lined or concentric butterfly valves. The main weak point of concentric butterfly valves is their liner, which can be damaged by particles, frequent cycling (opening and closing) of the valve or by operating the valve against high differential pressure.

5.3.1.1.5 Modular Valves

Modular valves, also called combination valves or double block and bleed (DBB) valves, are typically made of two ball valves with a needle valve in between inside one body, as illustrated in Figure 5.43. Having all three valves inside a single body rather than having three separate valves has various advantages, such as saving cost, weight and space. The main purpose of modular valves is to provide double isolation by closing the two ball valves. Double isolation may be required in high-pressure class piping and/or aggressive fluid to improve safety and reliability through redundancy. Using two blocks for isolation instead of one increases the cost, weight and required space of the valve and the complexity of the valve design. However, using an extra valve in a modular valve rather than a single valve for isolation assures that even if one valve fails, the other valve can provide fluid isolation. The needle valve, which is located between the two ball valves, acts as a bleeder, meaning that it releases the fluid trapped between the two ball valves. Modular valves are very common in small sizes, i.e., <2″. The important point is that a modular valve that includes two valves cannot be used

A lever showing half open position

A lever indicating half open position

Ball valve 1

Ball valve 2

Needle Valve

FIGURE 5.43   Modular valve including two ball valves in half-open position and a needle between. (Courtesy: Shutterstock.)

FIGURE 5.44   Position of the ball in half-open position. (Courtesy: Shutterstock.)

for fluid control. It is easy to see that both ball valves integrated into one body as a modular valve shown in Figure 5.43 are in a half-open position from the position of their levers. Ball valves are open fully when the lever stands parallel to the valve body and connected piping. The ball valves are fully closed when the levers stand perpendicular to the valve body and the connected piping. When the lever of the ball valves stands in the middle point, the valve can be assumed to be in a half-open position. Figure 5.44 illustrates the position of the ball in a half-open position in a modular valve. Keeping a modular valve in a half-open position for fluid throttling is not recommended, as it can create wearing and erosion inside ball valves. Modular valves are a very common choice for installation between piping and instrumentation in high-pressure and/or toxic fluid. As an example, modular valves can be installed between the piping or pressure vessel and the pressure transmitter. The other application of modular valves is to provide double isolation between high-pressure piping and chemical injection lines. As an example, anti-foam as a sort of chemical can be injected into a piping system connected to a separator to prevent foam from forming in the separator. Foam formation is known as a major separator operation problem. A separator is a pressure vessel used to separate produced oil, water and gas from the reservoir and the drilled well. Modular valves may be made by two plugs, two gates or two butterfly valves, rather than by two ball valves and a needle valve between. A needle valve is typically made in small sizes, such as two inches and smaller, for flow regulation, and is explained further in this chapter.

5.3.1.2 Fluid Control Valves

The amount of flow passing through the piping should be controlled in some cases. Flow control can adjust essential process variables, such as pressure, temperature and fluid level. In fact, valves for fluid flow regulation are chosen to regulate the flow by changing the fluid passage and keep the essential process variables as close as possible to the desired set point. As discussed above, butterfly valves can be used for fluid control. Additionally, globe, axial control, needle and choke valves are suitable for fluid regulation in topside offshore.

5.3.1.2.1 *Globe Valves*

Unlike ball, plug and gate valves, which are used to stop and start fluid flow in piping systems, globe valves are used for flow regulation. Although the American Petroleum Institute (API) Recommended Practice (RP) 615 valve selection guide states that globe valves may be used for blocking the fluid, the recommendation of the author according to industrial experience is to avoid selecting a globe valve for stop or start of the fluid. However, if a globe valve is required to provide 100% flow passage during complete opening or 0% flow passage during complete closing as part of the fluid control process, then a globe valve can be chosen. Globe valves typically have two main types of design; one is a T-pattern or standard design, and the other is a Y-pattern design. Figure 5.45 illustrates a globe valve including a part list. A globe valve with an actuator for automatic operation is called a *control valve*, illustrated in Figure 5.46.

Looking at the globe valve in Figure 5.45, the body is the main pressure-containing part that is connected to the bonnet or cover by means of bolts and nuts, and a gasket is placed between the body and bonnet for sealing purposes. The disk or plug sits on the seat in the closed position. A disk swivel nut or disk nut is used to connect the stem to the plug or disk. The rotation of the stem in rotating stem globe valves is not transferred to the disk in case of using a disk nut. In fact, using a disk nut assures that the stem and disk are not firmly connected, so the disk is always placed on the seats in a correct position without any rotation of the plug. The back seat is considered the primary valve stem sealing before packing.

The valve shown in the figure is in the closed position. To open the valve, the operator moves the handwheel counter-clockwise and the operator force is

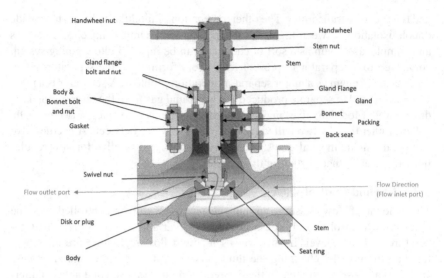

FIGURE 5.45   Globe valve with parts list. (Courtesy: Shutterstock.)

FIGURE 5.46   Control valve including a globe valve with diaphragm actuator. (Courtesy: Shutterstock.)

FIGURE 5.47　Cavitation on a metal surface. (Courtesy: Shutterstock.)

transferred to the stem through the stem nut. The stem and connected disk move upward and the valve opens. Thus, the fluid passes through the valve as per the route highlighted with dotted lines inside the valve. The fluid enters the valve from the right side, and after two 90° rotations in the middle of the valve close to the seat and plug, the fluid will leave the valve. A globe valve is not a bidirectional valve, meaning that it has a flow direction coming from under the disk, seat and stem area, as shown in the figure. The two 90° rotations of the fluid cause a significant pressure drop inside the valve. The significant pressure drops cause vaporization of the bubbles from the liquid fluid service; this is called *flashing*. The bubbles will recover their pressure after passing through the pressure drop area close to the disk and then burst. The burst of bubbles, called *cavitation*, is known as one of the major operational problems in tee-pattern or standard globe valves. Cavitation, which can be considered a type of erosion-corrosion, damages the body, seat, plug and stem of standard globe valves significantly. The other negative outcomes of cavitation are noise and vibration. Figure 5.47 illustrates cavitation corrosion, which is typically in the form of irregular pits on a metal surface.

There are some cavitation mitigation approaches, such as larger stem diameter, stronger connection of stem and plug as well as applying hard-faced materials such as stellite 6 or 21 to the seat and plug of the valve. Stellite is a type of cobalt and chromium alloy used for wear resistance on the valves' internals, such as the seat and plug. From an operational point of view, it is not recommended to keep the valve <20% open, as this can cause severe wearing and erosion, which can intensify the cavitation effect. The other solution is to change the type of valve to either a Y-type globe or an axial control valve. A Y-type or Y-pattern globe valve, as illustrated in Figures 5.48 and 5.49, reduces the pressure drop inside the valve as the rotation of flow inside the valve can be reduced to a maximum of two times and the flow direction change occurs at a maximum of 45°

FIGURE 5.48   Y-type globe valve from outside. (Courtesy: Shutterstock.)

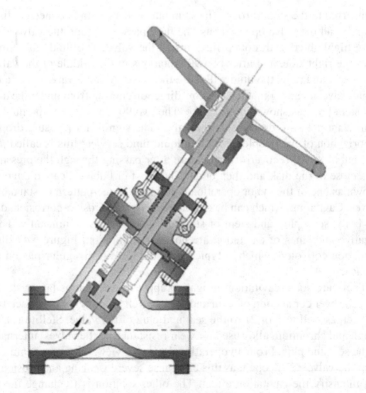

FIGURE 5.49   Y-type globe valve internals. (Courtesy: Hard Hat Engineering.)

rather than 90° in each rotation. Reduction of pressure drop value results in less cavitation.

5.3.1.2.2 Axial Valves

Axial control valves, as illustrated in Figure 5.50, are known as the best valves for flow control application compared to globe valves. The main characteristic of an axial control valve is a smooth and streamlined flow path inside the valve, which provides turbulence-free flow without operational problems such as vibration and noise. The pressure drop of the valve is low, which prevents cavitation and flow loss inside the valve. The other advantage of this valve is its compact and light design, which is an advantage in the offshore industry. It is very important to save weight and space in the offshore industry since there is limited space on the platforms. In addition, the platforms have limited load resistance capacity. The disk of axial control valves is light, and the distance between the disk and seat is short, meaning that the valve is suitable for fast opening and closing. In addition, the disk does not slam to the seat, so this type of valve has a non-slamming effect. The axial valve in Figure 5.50 is in the open position. To close the valve, the stem of the valve should be moved downward to push the disk forward; eventually, the disk will sit on the seat and the valve will be closed to stop the fluid. However, the stem movement can be adjusted to keep the disk in an intermediate position to control the flow of fluid passing through the valve. The other important component inside the valve is a cage installed between the disk and seat that affects the flow characteristic of the valve.

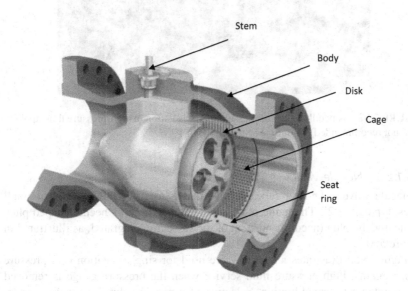

FIGURE 5.50 Axial control valve. (Courtesy: Mokveld Engineering.)

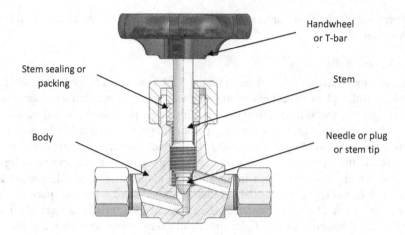

FIGURE 5.51   A needle valve with integrated stem and plug. (Photograph by author.)

FIGURE 5.52   A needle valve for pressure gauge isolation in high-pressure fluid application (not recommended). (Courtesy: Shutterstock.)

5.3.1.2.3 Needle Valves

A needle valve is another type of valve that is used for flow control in small sizes, typically <2″. The name of this valve comes from its needle-shaped plug. Typically, the plug (needle) and stem of the valve are integrated, as illustrated in Figure 5.51.

Figure 5.52 illustrates a needle valve used for single isolation of a pressure gauge against high-pressure fluid service when the pressure gauge is removed for maintenance or calibration. Selection of a needle valve for isolation, especially for high-pressure class services, is not recommended according to industry practices. A modular valve is the best and safest choice of valve for the isolation of high-pressure class fluid services.

5.3.1.2.4 Choke Valves

A choke valve is another type of control valve that is installed on the production wellhead to control the flow being produced from the well. In addition, a choke valve can be used to stop the production if something goes wrong downstream. Downstream refers to the piping and separator that are located after the choke valve. Figure 5.53a shows a choke valve on the right side of some Christmas tree valves. As explained earlier in this chapter, wellhead valves are all TCG valves. The right side of the picture shows a choke valve that is installed on a Christmas tree. The fluid enters the choke valve from the bottom part inside the vertical line; after a 90-degree rotation inside the valve, the fluid exits the valve to the horizontal line shown on the right-hand side of the choke valve in Figure 5.53. Choke valves are at high risk of operational problems such as erosion and cavitation because of the 90° rotation of the fluid inside the valve and the resulting pressure drop. The choke valve on the right side of Figure 5.53 has a stem and handwheel. The handwheel can be operated to move the stem and connected plug upward or downward to adjust the flow passage. In general, rotating the handwheel counter-clockwise moves the stem and connected plug upward, which increases the opening and allows for more flow to pass through the valve. Clockwise rotation of the handwheel has the opposite effect and closes the valve. Choke valves can be automatically operated by an actuator; in that case, there is no handwheel on the valve.

5.3.1.3 Non-Return Valves

Non-return valves, also called check valves, are designed and used to prevent backflow or reverse flow of liquid or gaseous media. Check valves have a flow

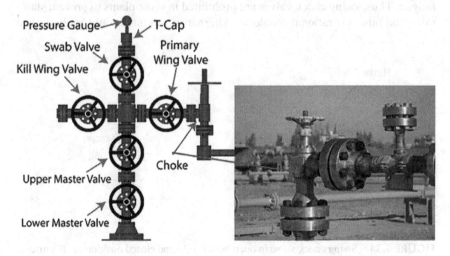

FIGURE 5.53 Choke valve on a wellhead. (Courtesy: Shutterstock.)

direction marked on the body of the valve showing the direction of flow in order to keep the valve open and allow fluid to pass through it. In fact, check valves do not have any means of operation; they open by means of fluid pressure. If the fluid stops and/or returns back, the valve closes and prevents the flow from moving in the opposite direction. In short, check valves are automatic valves that open with forward flow and close against reverse flow. There are different types of check valves, such as piston, swing, dual plate and axial. Different types of check valves for topside offshore are explained below.

5.3.1.3.1 Swing Check Valve

A swing check valve contains a disk that swings around a hinge. The hinge is connected to the body of the valve through a pin or shaft that passes inside the upper part of the hinge. A swing check valve is illustrated in the closed position on the left side of Figure 5.54. When the valve closes and the fluid returns to the upstream side of (before) the disk, it is prevented from flowing back by the closed disk. When the disk is closed, it is completely placed on the seat. The right-hand side of the figure shows that the fluid is moving from upstream of the valve on the right-hand side and is pushing the disk to the upward position so the fluid can pass through the valve. When the fluid stops, its weight and gravity force the disk back to its initial condition on the seat to close the valve. Thus, two conditions keep the disk tight to the seat: one is gravity and the other is the reverse flow. It is important to know that these two parameters could cause the disk to slam the seat when shutting down. Slamming is one of the major operational problems of swing check valves. Slamming has various negative consequences, such as damage to the disk and wearing of the hinge pin. In addition, slamming can cause pressure wave build and water hammering in the connected piping systems. Water hammering can damage piping, valves and instrumentation, and can cause a high amount of noise and acoustic fatigue. Thus, swing check valves are prohibited in some plants to prevent slamming and other operational problems. Alternative solutions could be dual-plate

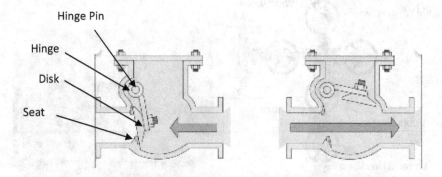

FIGURE 5.54 Swing check valve in open position (b) and closed position (a). (Courtesy: Shutterstock.)

Hinge

Disk

FIGURE 5.55   Swing check valve in closed position. (Courtesy: Shutterstock.)

and axial check valves. Dual-plate check valves cause only low to moderate slamming, and axial check valves are considered non-slamming. Regarding the cost of the valves, swing check valves are cheaper than dual plate and axial flow check valves in terms of capital cost or the cost of purchasing the valve. However, swing check valves can increase operational costs and thus the total cost of the valve compared to the two other choices. Axial valves are the most expensive check valves (Figure 5.55).

5.3.1.3.2   Dual Plate Check Valve

A dual-plate check valve contains double spring-loaded disks, as illustrated in Figure 5.56. How can such a design reduce water hammering significantly? The first reason is that the valve disks are not closed by gravity forces. Instead, spring force pushes the disks back to their initial state in the closed position. The second reason is that dual-plate check valves have two disks instead of one, so the weight of one disk is distributed onto two disks, which can lessen the slamming effect. Dual-plate check valves open by means of fluid pressure. When the fluid stops, the spring force is coupled with the force of the reverse flow and closes the valve. In fact, the valve is closed when the fluid rate decreases until the spring force overcomes the fluid pressure and keeps the plates closed. Check valves are used after pumps and compressors to protect this expensive equipment from backflow. Flow reversal from the discharge of the pumps and compressors toward the suction can damage these valuable facilities. Although dual-plate check valves can provide low to moderate slamming effect, they are not recommended for installation after pumps and compressors. The most suitable choice of valve to install after pumps and compressors is an axial check valve, which has a non-slamming characteristic. Axial check valves are explained in the next section.

FIGURE 5.56 Dual-plate check valve with spring-loaded double disks. (Photograph by author.)

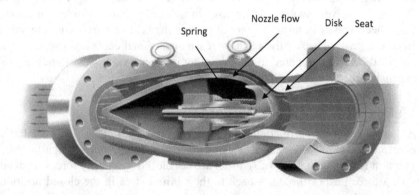

FIGURE 5.57 Non-slam check valve and the flow of fluid inside the valve. (Photograph by author.)

5.3.1.3.3 Axial Check Valve

Axial check valves, also called nozzle check or axial flow check valves, are very suitable valves for non-return fluid purposes due to their non-slamming and fast-closing characteristics. The name of the valve is taken from its internal structure (see Figure 5.57). As illustrated in the figure, there is a nozzle inside the valve. The figure indicates very smooth flow without any turbulence inside the valve. Pressure drop and flow loss in axial flow check valves are less than in swing and dual-plate check valves. The other advantage of this valve is its fast opening and closing, facilitated by the very short distance between its disk and seat, and its low-mass disk, which can move very quickly between open and closed positions. In fact, it is these two characteristics that prevent axial check valves from slamming. Nozzle check valves are the best choice of valves to install after pumps and compressors. Pumps and compressors are facilities that are used to pressurize the liquids and gasses in a piping system, respectively. Many reasons make axial check valves the best choice for installation after pumps and compressors; axial check valves can be closed very rapidly to prevent any backflow from passing through the valve and entering the upstream facilities, i.e., the pumps or compressors located before the valve. If the check valve located after the pumps and compressors does not operate properly and does not close quickly, the back or reverse flow enters the pumps and compressors and can cause damage. The other advantage of axial flow check valves for installation after pumps and compressors is that they are very resistant to vibration and have a robust design. It is important to bear in mind that pumps and compressors transmit a high amount of load and vibration to the connected piping systems, including the valves. Axial check valves require only a very small amount of maintenance, or even no maintenance, after years of operation due to their robust design. The amount of wearing and erosion is also very low compared to other types of check valves.

Axial flow check valves, like other check valves, have a flow direction. Refer to Figure 5.57; the fluid direction for opening the valve is from right to left. The working principle of axial check valves is as follows: when fluid enters the valve, the fluid pressure exceeds the spring force located behind the disk and pushes the disk back to open the valve. The flow passes through a narrow area located around the nozzle flow. When the pump or compressor that pressurizes the fluid in the piping system shuts down, the fluid pressure entering the valve is reduced and the spring force overcomes the fluid pressure and closes the valve.

5.3.1.3.4   Piston Check Valve

Swing and dual-plate check valves are not typically selected for small piping in sizes of 2″ and less because they create significant pressure drop inside the piping. The alternative check valve choice for piping in small sizes is a piston check valve. Figure 5.58 illustrates a piston check valve and its internal components: stem, spring, disk or plug and seat. A piston check valve opens when the fluid pressure and flow enter from the left side of the valve from under the disk. The fluid pressure overcomes the spring force and pushes the disk and connected stem upward,

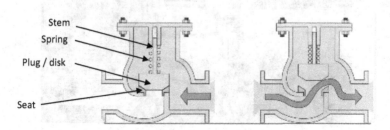

FIGURE 5.58   Piston check valve. (Courtesy: Shutterstock.)

opens the valve and passes through the valve as illustrated on the right-hand side of Figure 5.58. When the fluid pressure lessens, the disk moves down because of its weight and spring force, so the disk sits on the seat and closes the valve. As illustrated on the left side of the figure, the check valve remains closed when the fluid enters the valve from the right side, which is called reverse flow.

5.3.1.4 Safety Valves

5.3.1.4.1 Pressure Safety Valves

PSVs, as the name implies, have a safety function. The main aim is to install this type of valve on overpressure equipment or piping to prevent overpressure scenarios in the plant. PSVs release overpressure gasses from equipment in order to avoid overpressurizing and potential process safety incidents and to protect human life, property and the environment. An overpressure scenario refers to a condition that would cause a pressure increase in the piping or pressure equipment beyond the specific design pressure or maximum allowable working pressure. A PSV is considered the last line of defense for overpressure prevention. This means that other overpressure safety devices, such as emergency shutdown valves, should prevent overpressure scenarios before any requirement for a safety valve to act. Figure 5.59 illustrates a PSV inside a power plant; its spring is visible in the picture.

Figure 5.60 illustrates a conventional spring-loaded pressure relief valve. Like check valves, PSVs do not have any means of operation, such as an actuator, handwheel or gearbox. Instead, PSVs work automatically in response to increasing or decreasing fluid pressure inside the connected piping. The valve shown in the figure has a body and bonnet; these are pressure-containing parts, meaning that leakage from the body and bonnet leads to emission to the environment. The flow

FIGURE 5.59 Pressure safety valve in a power plant. (Courtesy: Shutterstock.)

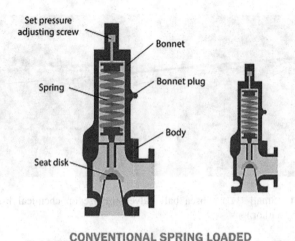

CONVENTIONAL SPRING LOADED
PRESSURE RELIEF VALVE

FIGURE 5.60   Conventional spring-loaded pressure relief valve. (Courtesy: Shutterstock.)

direction is from the bottom of the valve under the disk or seat disk. In the event of an overpressure scenario, the fluid pressure exceeds the spring load located above the disk, the disk is moved upward by the fluid pressure and the fluid rotates 90° and exits from the right side of the valve. When the pressure inside the connected piping system drops back to a normal pressure condition, then the spring force exceeds the fluid pressure inside the piping, the spring pushes the disk down, the disk sits on the seat and the valve closes. In a closed position, as illustrated in the figure, the disk is sitting on the seat area and preventing the passage of fluid through the valve.

5.3.2   Valves for Subsea

Valves that are used in the subsea oil and gas industry include ball, TCG (slab or expanding), modular, axial, needle, choke, swing check and axial check valves. The concept of operation and structure of subsea and topside valves are more or less the same. However, manually operated valves in topside and subsea could differ. Subsea ball valves are typically used on manifold headers in size ranges of 8″ and above. In addition, ball valves can be selected in sizes of <2″ in SDUs for chemicals and hydraulic fluid-containing piping systems (see Figure 5.61).

Ball valves may also be used for subsea isolation valves (SSIVs). An SSIV is a normally open actuated ball or axial valve that is used to create a safety barrier to protect the topside platform and its personnel from the release of hydrocarbon. The valve could be installed as a standalone module or installed inside a PLEM (see Figure 5.62); it is connected to the flowline (pipeline). In the event of leakage or release of hydrocarbon in the flowline or riser, the valve should be closed, i.e., in emergency mode. Since very fast operation is essential for an SSIV, this valve

½" subsea ball valves

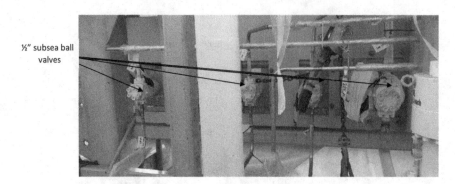

FIGURE 5.61 Small (½") subsea ball valves for use on chemical lines in SDUs. (Photograph by author.)

SSIV ball valve

PLEM structure

FIGURE 5.62 SSIV valve in a PLEM during subsea installation. (Courtesy: Shutterstock.)

is typically selected either in a ball or axial on/off valve. As mentioned above, a ball valve is a quarter-turn valve capable of fast operation. An axial on/off valve, like an axial control valve, is also capable of fast operation because of its low-mass disk and the short distance between the disk and seat. An axial on/off valve, unlike an axial control valve, is just used for on/off application and not flow regulation. An SSIV valve is typically actuated, so its actuator is connected to an umbilical.

Subsea TCG valves are located on subsea trees and the branch lines of subsea manifolds. Figure 5.63 illustrates a 7 1/16" subsea gate valve inside a manifold. Subsea needle valves can be located on small (e.g., 1") chemical lines. Subsea modular valves (see Figure 5.64) are alternative valves on chemical injection lines that provide double isolation between the chemical and the process line. Subsea choke valves are located on the subsea well production line to regulate the produced flow

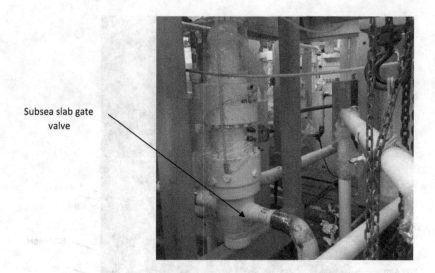

Subsea slab gate
valve

FIGURE 5.63   Subsea slab gate valve in a subsea manifold. (Photograph by author.)

FIGURE 5.64   Subsea modular valves. (Photograph by author.)

according to the same concept mentioned above regarding topside choke valves. Axial valves are typically small in size (e.g., 2″ or ¾″) and are installed on chemical injection lines to prevent backflow of the main process fluid into the chemical injection line. In fact, axial valves on chemical injection lines open when chemicals are injected into the process line, and close when the chemical injection stops. Subsea swing check valves can be installed on pipelines in relatively large sizes.

Although subsea valves may have a handwheel as a means of operation, like topside valves, that can be operated by a diver (see Figure 5.65a) in shallow water

FIGURE 5.65   A diver (a) and an underwater ROV (b) subsea. (Courtesy: Shutterstock.)

depths of up to 100m, that is a rare case. The subsea oil and gas industry is moving toward deeper areas of seas and oceans in depths between 1 and 3 km—or even deeper—where a diver cannot go. Instead, remotely operated underwater vehicles (ROVs) (see Figure 5.65b) are used to manually operate valves installed at depths greater than roughly 100m, where divers cannot swim. An ROV is a kind of underwater robot typically connected to a ship by a series of cables. The cables transmit command and control signals from the operator to the ROV. Thus, the navigation of the ROV is performed by an operator on board the ship. The transmission of signals and electrical power between the ROV and the surface takes place by means of a *tether*. An ROV typically has some *lights* for illumination and is equipped with a *camera* to take pictures if necessary. An ROV gets its power for movement from the electrical or hydraulic power source connected to its *thrusters*. All ROV components are located inside a *frame*. An ROV has arms with fingers at the end capable of holding a torque tool. A torque tool is a device that is placed inside the ROV bucket or paddle to operate the valve. ROV buckets installed on the top of modular valves are shown in orange in Figure 5.64.

5.4 INDUSTRIAL ACTUATORS FOR THE OFFSHORE INDUSTRY

5.4.1 TOPSIDE ACTUATORS

An actuator is a type of mechanical or electrical component that is installed on a valve for automatic operation, as explained briefly in the introduction to this

FIGURE 5.66 Actuated ball valve on an offshore platform. (Courtesy: Shutterstock.)

chapter. Using an actuator for a valve means that the actuator can automate the valve without any need for human involvement or interaction. Figure 5.66 illustrates a ball valve with a pneumatic actuator on an offshore platform. There is a control panel box installed beside the actuator to control and command the airflow to the actuator. Actuators are categorized according to their source of power as well as their type of motion (rotary or linear). The source of power for topside actuators could be air, hydraulic oil, electricity, gas, a mixture of hydraulic oil and gas or a combination of hydraulic oil and electricity. Considering the source of power, actuators for topside could be hydraulic, pneumatic, electrical, electro-hydraulic, gas-over-oil or direct gas. The other way to categorize actuators is based on the motion provided for the valve, which could be linear or rotary. The next section provides information about each type of actuator.

5.4.1.1 Linear Actuators

5.4.1.1.1 Diaphragm Actuators

Diaphragm actuators typically work with air pressure, so they are categorized as a type of pneumatic actuator. Diaphragm actuators are a very common choice of automation for control valves. Diaphragm actuators are used in all sectors of the oil and gas industry except subsea. Figure 5.67 illustrates a diaphragm actuator installed on a control valve.

The upper diaphragm casing should be removed to provide access to the internal components. Figure 5.68 illustrates the control valve internals, including the diaphragm, a plate and connected stem and a spring. The air pressure could be applied to the top of the diaphragm, as illustrated in Figure 5.68; this is known as a *direct-acting* diaphragm actuator. The *reverse-acting* diaphragm actuator on the opposite side is a type of diaphragm actuator in which the air enters the actuator from the bottom. The main working principle of a direct-acting diaphragm actuator is that the air is applied to the top of the diaphragm and pushes the stem downward. Diaphragm actuators provide linear motion for control valve operation. Direct-acting actuators could be opened or closed by means of air pressure.

FIGURE 5.67 Control valve with a diaphragm actuator. (Courtesy: Shutterstock.)

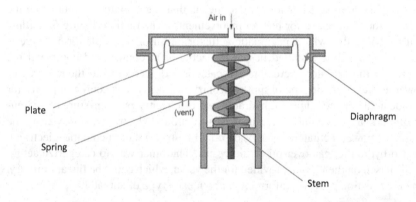

FIGURE 5.68 Direct-acting diaphragm actuator internals. (Courtesy: Instrumentation Tools.)

But the most common type of direct-acting actuator for a control valve is one that closes by means of the pressure of the air above the diaphragm. In the event that a direct-acting actuator loses air, the spring pushes the plate and the valve internal upward and the valve opens. An actuator that closes with air and opens with spring force has a fail-safe open function.

The main working principle of a reverse-acting actuator is that the air pressure enters the actuator from the bottom of the diaphragm and pushes the diaphragm up. The control valve is typically opened by means of air pressure in a reverse-acting actuator. If the air supply is cut off because of a failure in the process system, such as an overpressure scenario, the spring force closes the valve. In a reverse-acting actuator, spring is located above the diaphragm, opposite from the air entrance to the actuator. Thus, reverse-acting actuated control valves mainly have a fail-safe closed function.

Hydraulic
linear piston
actuator

Through
conduit gate
valve

FIGURE 5.69   Hydraulic linear piston actuator on a TCG valve. (Photograph by author.)

5.4.1.1.2   Piston Actuators

Linear piston actuators can work with either air or hydraulic sources of energy, meaning that they could be either pneumatic or hydraulic. The selection between hydraulic or pneumatic depends on different parameters, such as valve type, size, pressure class, etc. Hydraulic fluid can produce a higher force for the operation of actuators and valves than pneumatic power. Therefore, for a type of valve like a ball valve, which requires more force for operation, especially in large sizes and high-pressure classes, a hydraulic actuator could be a better choice. Linear piston actuators create motion in a straight line and are very common for TCG valves (see Figure 5.69). Linear piston actuators are typically installed on valves vertically.

Linear actuators contain a piston inside the cylinder. The movement of the piston provides the required motion for valve operation. The source of the piston movement could be a spring, air or hydraulic oil. There are two types of design with regard to the source(s) of power operating an actuator: single-acting and double-acting. Single-acting actuators, also called spring return actuators, are operated by a spring from one side and by air or hydraulic fluid from the other side. In this case, the spring of the actuator can move the valve to either open or closed position. If the spring closes the valve, the actuator has a fail-close or fail-safe closed function. Alternatively, if the spring opens the valve, the actuator has a fail-safe open function. Fail-safe closed piston-type actuators are more common in the offshore oil and gas industry. Linear piston-type actuators could be opened and closed by hydraulic oil. Piston actuators could be opened and closed by hydraulic oil from both sides, as illustrated in Figure 5.70. In this case, the actuator is called double-acting. Double-acting actuators remain in the last position in case of hydraulic oil supply stoppage, meaning that they have a fail-as-is

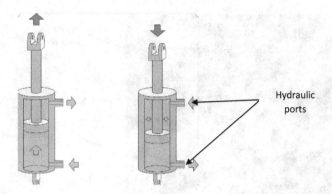

Hydraulic ports

FIGURE 5.70 Double-acting piston hydraulic actuator with linear motion. (Courtesy: Shutterstock.)

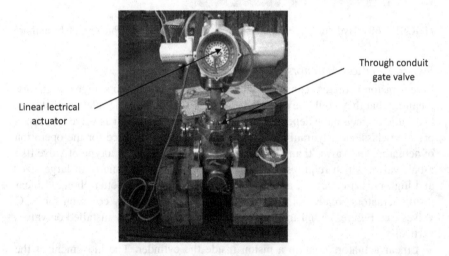

Through conduit gate valve

Linear lectrical actuator

FIGURE 5.71 Linear electrical actuator on a TCG valve. (Photograph by author.)

function. Thus, double-acting actuators do not have a fail-safe closed or fail-safe open function to return to either closed or open positions in the event of power supply stoppage.

5.4.1.1.3 Electrical Actuators

Electrical actuators, also called electrical motors, can provide linear motion. Linear electrical actuators can be installed on the top of TCG valves, as illustrated in Figure 5.71. In addition, electrical actuators can generate rotary motion applicable for quarter-turn valves like ball, plug and butterfly valves. Electrical actuators have a lower speed of operation compared to both pneumatic and hydraulic actuators. The other important characteristic of electrical actuators is that they typically stay in the last position upon stoppage of the motor, meaning that they have

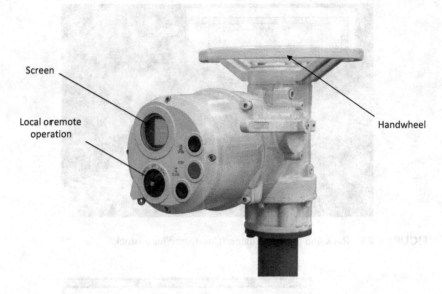

Screen

Local or remote operation

Handwheel

FIGURE 5.72   Electrical actuator. (Courtesy: Shutterstock.)

a fail-as-is function. Electrical actuators could be coupled with gearboxes in order to produce higher force for the operation of the valves. Figure 5.72 illustrates an electrical actuator with three buttons: one for opening, one for closing and one for stopping the actuator. These three buttons work when an operator is operating the actuator locally. Local operation means that the operator is standing next to the actuator and is not typically permitted without a signal from the control room. A control room is a large operation room with facilities used to monitor and control the plant. There is a big button on electrical actuators that can be used to switch between local and remote modes of operation. Remote operation of the actuator can be performed from the control room. There is a handwheel on an electrical actuator for manual operation. The screen on an electrical actuator can show the percentage of valve opening and the amount of force produced by the actuator.

5.4.1.2   Rotary Actuators

Unlike linear actuators, rotary actuators provide rotary motion for the operation of valves such as ball, butterfly and plug valves. Rotary type actuators that are applicable for topside could be scotch and yoke, rack and pinion, rotary electrical, electro-hydraulic, direct gas and gas-over-oil.

5.4.1.2.1   Rack and Pinion Actuators

Rack and pinion is a type of rotary actuator that can work with air or hydraulic oil. As the name suggests, it has two gears—rack and pinion—as illustrated in Figure 5.73. The rack has a linear gear movement that moves against the pinion's circular gear. Rack and pinion actuators can have either a fail-safe or fail-as-is

FIGURE 5.73   Rack and pinion actuator. (Courtesy: Shutterstock.)

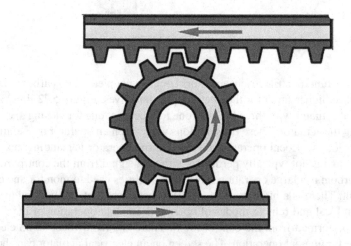

FIGURE 5.74   Rack and pinion with double rack. (Courtesy: Shutterstock.)

function. Fail-safe rack and pinion actuators have a spring on one side and air or hydraulic oil pressure on the other side for movement of the rack against the pinion. Fail-as-is actuators are double-acting, which means that the rack moves back and forth against the pinion by means of hydraulic oil from both sides. The rack and pinion mechanism in the figure has a single rack. A double or symmetrical rack could be used to provide higher force and more importantly, a pressure balance feature, as illustrated in Figure 5.74.

5.4.1.2.2   Scotch and Yoke Actuators
Scotch and yoke is another type of rotary actuator, suitable for quarter-turn valves such as ball and butterfly. This type of actuator can work with hydraulic power or air

systems and could be either single-acting (spring return) or double-acting. Figure 5.75 illustrates a spring return scotch and yoke actuator that converts linear or reciprocating motion to the rotary or quarter-turn motion. The shaft in the middle of the actuator is directly connected to the yoke through a slot or pin. The air or hydraulic fluid enters the actuator from the left side and moves the shaft forward. The linear motion of the shaft is transferred to the yoke through the slot or pin. The yoke and connected shaft then rotates 90° and moves the valve between open and closed positions.

Scotch and yoke actuators could be double-acting (springless), i.e., pressurized by air or oil from both sides. A scotch and yoke actuator with a single-acting or fail-safe mode of operation has a housing that contains the yoke mechanism and a pressure cylinder that contains the piston and spring enclosure.

5.4.1.2.3 Rotary Electrical Actuators

Electrical actuators can provide rotary motion for quarter-turn valves such as ball and butterfly valves. Figure 5.76 illustrates a 38″ ball valve automated with an

Slot or pin

Hydraulic
or air

Yoke

Piston

FIGURE 5.75 Scotch and yoke actuator. (Courtesy: Shutterstock.)

Rotary electrical
actuator

FIGURE 5.76 Electrical actuator on a large 38″ ball valve. (Photograph by author.)

electrical actuator. The aim of using an electrical actuator for such a large valve is to provide ease of operation for a human operator. If the valve were supplied with a handwheel and gearbox for manual operation, it would take a long time and many turns of the handwheel for an operator to open and close such a large valve.

5.4.1.2.4 Rotary Electro-Hydraulic Actuators

Electro-hydraulic actuators are a kind of hydraulic actuator that works with electrical motors; in fact, an electro-hydraulic actuator includes two connected and integrated devices; one electrical motor and one hydraulic actuator. The main advantage of an electro-hydraulic system is that it eliminates the required hydraulic oil distribution system including the pumps and tubing. The electrical motor can be used for providing power to move the hydraulic oil, or the hydraulic pump can be combined with the electrical and hydraulic actuator. The other advantage of electro-hydraulic actuators compared to hydraulic actuators is higher safety and reliability. Unlike electrical actuators, electro-hydraulic actuators can provide a fail-safe mode of operation. Figure 5.77 illustrates an electro-hydraulic actuator with a hydraulic actuator and an electrical motor that are integrated together.

5.4.1.2.5 Rotary Direct Gas Actuators

Direct gas actuators could be linear or rotary. Linear direct gas actuators are suitable for gate valves, while rotary direct gas actuators are suitable for quarter-turn valves like ball valves. Gas is the source of fluid for operating such actuators. The source of natural gas for the operation of the actuator could be the piping or pipeline connected to the valve. Direct gas actuators are only available in a double-acting design. Figure 5.78 illustrates a double-acting rotary direct gas actuator for a ball valve. The actuator is opened and closed by means of gas pressure. The main disadvantage of this type of actuator is that the gas is a source of fugitive emission, which is its most essential negative impact; any leakage of gas from the actuator is undesirable and can cause environmental pollution.

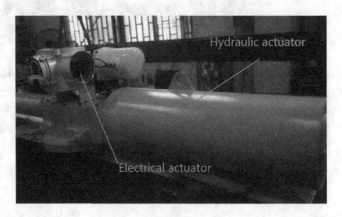

FIGURE 5.77 Electro-hydraulic actuator. (Photograph by author.)

FIGURE 5.78   Direct-gas double-acting rotary actuator.

FIGURE 5.79   Gas-over-oil actuator. (Photograph by author.)

5.4.1.2.6   *Rotary Gas-over-Oil Actuators*

This type of actuator uses gas from the pipeline to pressurize the hydraulic fluid
to move a double-acting hydraulic actuator. Using high-pressure gas eliminates
the requirement for using pumps and extra tubing to pressurize the hydraulic oil.
Figure 5.79 illustrates a gas-over-oil actuator used for a double-acting scotch yoke
actuator. As shown in the figure, the actuator has capsules that contain both oil
and gas. In the same manner, as rotary direct gas actuators, using gas inside a
gas-over-oil actuator could be a source of fugitive emission. Some valves installed
in remote areas, such as those located on pipelines, are actuated by natural gas
pressure and rotary gas-over-oil actuators.

5.4.2 SUBSEA ACTUATORS

Subsea actuators can be either hydraulic or electrical. Thus, air or gas are not used as sources of energy for subsea actuators. Subsea hydraulic actuators may be single-acting or double-acting, and subsea hydraulic actuators could be either linear or rotary. Linear actuators are mainly used for subsea gate valves, while rotary actuators are mostly applicable for subsea ball valves.

The subsea oil and gas industry is continuously moving toward more simple designs to save weight and space, implement digitalization and reduce both expenditure (CAPEX) and operational costs (OPEX). Moreover, keeping the environment clean by preventing emission and spillage is always a concern. It is important to bear in mind that the venture for oil and gas exploration and production is moving toward ever deeper and more remote areas of the seas and oceans. To achieve the desired aims of cost-reduction and maintaining a clean environment, an all-electrical control and actuation system is preferred over electro-hydraulic subsea distribution and control with hydraulic actuators.

The use of all-electrical actuators instead of electro-hydraulic actuators has many benefits, and the oil industry has been keen to move to the concept of all-electrical subsea systems for many years. Many research programs have been conducted to justify and evaluate all-electrical subsea systems. Shifting from hydraulic to electrical systems, which necessitates changing all subsea hydraulic actuators to electrical types, has many advantages that can be summarized as reducing cost, improving health, safety and environmental (HSE) protection and increasing functionality, reliability and flexibility.

5.5 CONCLUSION

Different types of valves and actuators for the offshore industry including both topside and subsea are discussed. Valves are categorized based on their applications like fluid stop or start, flow regulation, preventing fluid back and safety purposes. Actuators can be divided based on the source of energy like electrical, hydraulic, hydraulic-electrical and pneumatic.

5.6 QUESTIONS AND ANSWERS

1. Valves and actuators are located on which of the following structures or components in the offshore industry?
   A. Platforms and ships
   B. Subsea manifolds and trees
   C. Subsea umbilicals
   D. Subsea PLET and PLEM
   E. Subsea distribution units (SDU)
   Answer: Option A is correct since valves and actuators in the top-side oil and gas industry are located on a platform or ship. Option B is

correct as well, since most subsea valves and actuators are located on manifolds and trees. Option C is not correct, because subsea umbilicals are a mixture of hoses and cables that transfer hydraulic fluid and electricity and they do not have any valves or actuators. A subsea PLET, which is used to connect two pipes, does not have any valve, but a PLEM, as a type of manifold, normally does contain valves and actuators, so option D is partially correct. An SDU typically contains small valves, so option E is correct as well. In conclusion, options A, B and E are correct.

2. Which type of valve is proposed for fluid isolation in a clean oil service in an operating pressure of 60 bar and an operating temperature of 150°C? The size of the connected piping is 20″, and the pipe is horizontally oriented. It is an important requirement in this case that the height of the valve should be as short as possible, as per the piping designer's request, to avoid clashes.

    A. Gate valve
    B. Ball valve
    C. Globe valve
    D. Plug valve

    Answer: Option A is not the best choice for one main reason: gate valves (expanding, slab and wedge type) have high height, which is a concern in this case. Option C is not correct, since a globe valve is used for flow regulation and not for flow isolation. Option D is not correct either, since a plug valve is proposed for particle-containing services, whereas the fluid is clean in this case. Thus, option B, ball valve, is correct.

3. The main concern is to select a compact valve for a seawater piping system in pressure class 150 equal to 20 bar. Cost is another issue, meaning that the valve should not be very expensive. The required valve has an emergency shutdown safety-critical function and should be capable of fast closing in case of emergency to shut down the line. Which type of valve would you propose for such an application?

    A. Ball valve
    B. Concentric wafer butterfly valve
    C. Double offset wafer butterfly valve
    D. Triple-offset wafer butterfly valve

    Answer: A ball valve is more expensive than a butterfly valve; it is also bulkier and requires more space. Both cost and compact design are essential parameters in this example, meaning that using a ball valve compared to a butterfly valve would result in increasing both cost and required space. It is important to pay attention to the fluid service, which is water (a non-process service), so butterfly valves could be a good choice. Butterfly valves can be designed without a flange ending, which is known as wafer design, to save cost and space considerably. Figure 5.80 compares the face-to-face dimensions or length of

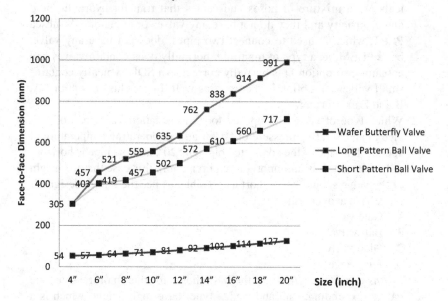

Face-to-face comparision of class 150 ball and wafer butterfly valves in sizes from 4" to 20"

FIGURE 5.80 Comparison between wafer-type butterfly and ball valve face-to-face dimensions (size range of 4"–20" in pressure class 150). (Photograph by author.)

wafer-type butterfly and ball valves in a size range of 4"–20" in class 150. On average, wafer butterfly valves in the given size range and pressure class are six times more compact than ball valves. Thus, option A is not correct. Concentric butterfly valves are not robust valves due to having a rubber liner that could be damaged, in which case the valve should be replaced. Thus, rubber-lined butterfly valves cannot meet the requirement of being selected for safety-critical functionality, and option B is not correct. Both double-offset and triple-offset butterfly valves are suitable for this application, but triple-offset butterfly valves are more expensive than double-offset, and since reducing the cost is a concern here, the correct option is a double-offset wafer butterfly valve. Therefore, option C is correct.

4. Cavitation is not tolerated at all from a valve to be used for flow regulation; which valve is proposed for such a case?

A. Butterfly valve
B. Globe valve
C. Axial control valve
D. Y-type globe valve

Answer: Cavitation can happen in both globe and butterfly valves. Y-type globe valves have less cavitation risk, but axial control valves are cavitation-free. Thus, option C is correct.

5. Which sentences are correct about check valves?
   A. A dual plate check valve is a suitable choice of valve for sizes of 2″ and less.
   B. An axial check valve can provide less pressure loss compared to dual plate and swing check valves.
   C. Swing check valves are the least expensive valves regarding the initial or capital cost in comparison with dual plate and axial check valves.
   D. Reducing the fluid pressure upstream of a dual plate check valve causes the valve to be closed by the weight of the plates.

   Answer: Option A is not correct. Dual plate check valves can cause a considerable amount of pressure drop in sizes of 2″ and less, so they are not recommended. Piston check valves are suitable for small sizes as low as 2″ and less. Option B is completely correct, as axial check valves have the least pressure loss compared to swing and dual plate check valves because of the venturi effect inside axial valves. Option C is correct also; swing check valves have a lower initial cost compared to dual-plate and axial valves. Option D is partially correct, as reducing the fluid pressure upstream of a dual plate check valve causes the valve to be closed by spring force and not by the weight of the plates. Thus, options B and C are correct.

6. Which option gives all the essential characteristics of an axial check valve?
   A. Non-slamming, compact, light, slow opening and closing
   B. Smooth flow, high flow capacity, fast opening and closing
   C. Non-slamming, fast opening and closing, robust design, low-mass disk
   D. Medium slamming, short distance between the disk and seat, compact and light

   Answer: Option A is not the correct answer. Non-slamming is the main characteristic of an axial valve, but axial valves are not necessarily compact and light. Additionally, the opening and closing of axial check valves are fast. Option B is not completely correct either, as axial check valves do not have high flow capacity. The valve internals, such as the disk and flow nozzle, are located on the flow path, which reduces flow capacity. However, two other characteristics of check valves, which are smooth flow and fast opening and closing are correct. Option C is correct and addresses all the essential characteristics of check valves. Option D is not correct, because medium slamming and compact and light design are not characteristics of axial check valves. Thus, option C is the correct answer.

7. Which sentences are correct about subsea valves?
   A. Subsea ball valves are always large in size, i.e., 8″ or larger.
   B. Choke valves are typically selected for subsea manifolds.
   C. TCG valves are only selected for subsea trees.

D. Subsea needle valves are typically in small sizes <2″ and are used for chemical injection lines.

Answer: Option A is not correct because subsea ball valves can be selected in smaller sizes, such as ½″, ¾″ or 1″ in SDUs on chemical and hydraulic lines. A 8″ ball valves and above are located on manifolds. Option B is not correct, since choke valves are typically connected to trees located upstream of (before) the manifolds. Option C is not correct, because TCG valves are used on both trees and manifolds. Option D is the only correct answer.

8. Fill in the gaps with the correct words.

A 38″ ball valve in a high-pressure class of 2,500 with a fail-safe closed function requires a/an _____ actuator for operation. This type of actuator is categorized as _____. The other type of actuator, which provides a fail-as-is function, is _____. This type of actuator with fail-as-is function can provide both _____ and linear motions.

A. Electrical, linear, pneumatic, rotary
B. Hydraulic, rotary, electrical, rotary
C. Pneumatic, rotary, hydraulic, linear
D. Electrical, linear, electrical, linear

Answer: Electrical actuators typically cannot provide a fail-safe closed mode of operation. Since the actuated 38″ ball valve is fail-safe closed, then the remaining options for actuation are either pneumatic or hydraulic. A 38″ ball valve in pressure class 2,500 is a large valve in a high-pressure class, so it requires a large amount of force for operation; pneumatic actuators cannot generate the force required to operate such a large valve in such a high-pressure class. Thus, the correct actuator is hydraulic. A ball valve is a quarter-turn valve with rotary motion. The other type of actuator that provides a fail-as-is function is an electrical actuator. Electrical actuators provide both linear and rotary motion, so option B is the correct answer.

9. Which types of actuators are applicable for the subsea oil and gas industry?

A. Hydraulic and electrical actuators
B. Gas-over-oil and direct-gas actuators
C. Pneumatic and electrical actuators
D. None of the above

Answer: Hydraulic and electrical actuators are applicable for the subsea oil and gas industry, so option A is correct. Gas-over-oil and direct-gas actuators are not applicable for subsea, so option B is not correct. Option C is not completely correct, because pneumatic actuators are not applicable for subsea, but subsea electrical actuators do exist. Option D is not correct either. In conclusion, option A is correct.

10. Which sentences are correct regarding subsea all-electrical actuators?

A. All-electrical subsea actuators will be the future of subsea.

B. Shifting from the current subsea hydro-electrical system to an electrical system leads to higher cost.

C. Subsea electrical actuators have better reliability compared to hydraulic actuators.

D. All-electrical subsea actuators are a good alternative to subsea pneumatic actuators.

Answer: Option A is correct; all-electrical subsea actuators will be the future of subsea. Option B is wrong because all-electrical systems are cheaper than electro-hydraulic systems. Option C is correct, as subsea electrical actuators are more reliable than subsea hydraulic actuators. Option D is not correct, because pneumatic actuators are not used subsea. Thus, options A and C are correct.

BIBLIOGRAPHY

1. American Petroleum Institute (API) 6A (2018). *Specification for Wellhead and Tree Equipment*, 21st edition. API, Washington, DC.
2. American Petroleum Institute (API) RP 615 (2016). *Valve Selection Guide*, 2nd edition. API, Washington DC.
3. American Society of Mechanical Engineers (ASME) B16.34 (2020). *Valves— Flanged, Threaded, and Welding End*. ASME, New York.
4. International Organization of Standardization (ISO) 10423 (2009). *Petroleum and Natural Gas Industries—Drilling and Production Equipment—Wellhead and Christmas Tree Equipment*, 4th edition. ISO, Geneva, Switzerland.
5. Nesbitt, B. (2007). *Handbook of Valves and Actuators: Valves Manual International*, 1st edition. Elsevier: Oxford.
6. Sotoodeh, K. (2015). Axial flow nozzle check valves for pumps and compressors protection. *Valve World Magazine*, Vol. 20, No. 1, pp. 84–87.
7. Sotoodeh, K. (2018). Comparing dual plate and swing check valves and the importance of minimum flow for dual plate check valves. *American Journal of Industrial Engineering*, Vol. 5, No. 1, pp. 31–35.
8. Sotoodeh, K. (2018). Selecting a butterfly valve instead of a globe valve for fluid control in a utility service in the offshore industry. *American Journal of Mechanical Engineering*, Vol. 6, No. 1, pp. 27–31.
9. Sotoodeh, K. (2019). *Actuator Sizing and Selection*. Springer Nature Applied Science, Switzerland. doi: 10.1007/s42452-019-1248-z.
10. Sotoodeh, K. (2021). Safety and reliability improvements of valves and actuators for the offshore oil and gas industry through optimized design. University of Stavanger. PhD thesis, UiS. no. 573.
11. Sotoodeh, K. (2021). *A Practical Guide to Piping and Valves for the Oil and Gas Industry*, 1st edition. Gulf Professional Publishing, Elsevier Science, Oxford.
12. Sotoodeh, K. (2021). *Subsea Valves and Actuators for Oil and Gas Industry*, 1st edition. Gulf Professional Publishing, Elsevier Science, Oxford.
13. Sotoodeh, K. (2021). *Prevention of Actuator Emissions in the Oil and Gas Industry*, 1st edition. Gulf Professional Publishing, Elsevier Science, Oxford.
14. Sotoodeh, K. (2021). Analysis and failure prevention of nozzle check valves used for protection of rotating equipment due to wear and tear in the oil and gas industry. *Journal of Failure Analysis and Prevention*. Springer. doi: 10.1007/s11668-021-01162-2.

6 Piping, Valves and Actuator Offshore Coating Case Studies

6.1 INTRODUCTION

This chapter addresses and details the experiences and lessons learned from the coating of piping, valves and actuators in offshore projects, including both topside and subsea. Certain coating issues with regard to piping are also mentioned in this chapter. As discussed in Chapter 1, the offshore environment is very harsh and corrosive, due to its chloride-containing atmosphere and the presence of seawater. Thus, the usage of coating to prevent external corrosion is more common in offshore compared to onshore oil and gas plants, which are located away from the corrosive marine environment. Figure 6.1 illustrates a non-slam axial check valve in 22Cr duplex to be installed after a pump. The operating temperature of the check valve is higher than 100°C, so thermal-spray aluminum (TSA) coating, NORSOK system 2, is applied on the valve. However, two areas close to the end of the valve remain uncoated shortly after valve installation. As the figure clearly illustrates, the uncoated areas in duplex material are already corroded. The type of corrosion shown in the figure looks like general corrosion; it occurred after just 1 month of keeping the valve in the offshore environment during the installation process. This example shows that coating is an essential aspect of external corrosion protection in the offshore environment, and that lack of coating can lead to severe corrosion attack on valves and other components in a remarkably short period of time.

6.2 COATING APPLICATIONS FOR TOPSIDE VALVES AND ACTUATORS

6.2.1 TOPSIDE PIPING AND VALVE COATING CASE STUDIES

6.2.1.1 No Coating on Titanium and Nickel Aluminum Bronze Valve Bodies

Coating is widely applied to carbon steel and stainless steels in the offshore industry. However, valves used in corrosive seawater systems could be made of nickel aluminum bronze (NAB) and titanium; both NAB and titanium offer excellent resistance against corrosion types generated by seawater, such as crevice, pitting and chloride stress cracking corrosion (CSCC). More information about these

DOI: 10.1201/9781003255918-6

Uncoated areas

FIGURE 6.1 A non-slam check valve coated in TSA—the uncoated areas have been attacked by general corrosion. (Photograph by author.)

TABLE 6.1
Corrosion Resistance Comparison between NAB and Titanium

	General Corrosion	Pitting Corrosion	Crevice Corrosion
NAB	9	7	8
Titanium	10	10	10

types of corrosion is provided in Chapter 1. Titanium piping and valves are widely used in the Norwegian offshore industry in corrosive seawater services. Titanium is much more expensive compared to NAB, but it provides better corrosion resistance compared to NAB. Table 6.1 compares the corrosion resistance of NAB and titanium. The numbers in the table are given from 1 to 10, indicating that titanium provides the highest amount of corrosion resistance.

Butterfly valves are widely used in seawater services in the Norwegian offshore industry. Figure 6.2 shows two butterfly valves; the valve on the right in white color is in titanium, and the other in golden color is in NAB. As you can see in the figure, the butterfly valves in these two materials, titanium and NAB, remain uncoated, as these materials provide excellent corrosion resistance in chloride-containing environments and do not require coating.

6.2.1.2 Manual Valve Coating Standardization

The coating standardization discussed in this chapter is applicable to topside valves. Valves are categorized into manual and actuated, as explained in Chapter 5. Manual valves are categorized as bulk items, while each actuated valve

FIGURE 6.2 Titanium (a) and NAB (b) butterfly valves may be used offshore without the need for any coating for external corrosion protection. (Photograph by author.)

TABLE 6.2
Coating Systems for Topside Valves in an Offshore Project

Valve Body Material	Operation Conditions	Coating
Carbon steel	Operating temperature > 120°C	TSA
	Operating temperature ≤ 120°C	Inorganic zinc coating
Stainless steel 316	Operating temperature < 60°C	Not applicable
	Insulated and operating temperature < 60°C	Phenolic epoxy
22Cr duplex	Operating temperature > 100°C	TSA
	Operating temperature ≤ 100°C and no insulation	Not applicable
	Operating temperature < 100°C and insulated	Phenolic epoxy
25Cr duplex	Operating temperature ≤ 20°C and no insulation	Not applicable
25Cr duplex	Operating temperature ≤ 20°C and insulated	Phenolic epoxy
6MO	Operating temperature > 120°C	TSA
	Operating temperature ≤ 120°C and no insulation	Not applicable
	Operating temperature ≤ 120°C and insulated	Phenolic epoxy
Titanium		Not applicable
NAB		Not applicable

has a unique tag. The coating system for actuated valves is normally selected case by case, according to the substrate material and operating temperature, since each one has a unique tag. Table 6.2 provides the coating systems for various valve materials in different operating temperature conditions. It should be noted that 25Cr duplex is normally selected for use in an operating temperature at a maximum of 20°C for seawater service in the Norwegian offshore industry. However, if 25Cr super duplex is selected for a process service like oil and gas with an operating temperature higher than 110°C, then TSA coating is recommended. Titanium and NAB are highly resistant to the corrosive elements in the marine environment, so they do not require any coating, as discussed in the previous section.

The benefits of coating standardization are to avoid having different coating varieties on similar valves and to simplify the control and management of the valves. As an example, let's suppose that we have 30, 12″ manual ball valves in pressure class 150, meaning that they are suitable for 20 bar pressure, in carbon steel material, and that all of these valves are completely identical with the same bore or internal diameter and the same flanged end connection. However, 20 of these valves have an operating temperature of more than 120°C, and 10 have an operating temperature of <120°C. Therefore (refer to Table 6.2), 20 of these valves should be coated with TSA and 10 should be coated with inorganic zinc coating. One possible solution to better manage and control the valves and their coating is to standardize the coating of all 30 valves with TSA. However, it should be noted that coating ten of the valves with TSA instead of inorganic zinc coating increases the cost of the coating. It would not be a correct decision to standardize the coating of all 30 pieces of the manual butterfly valves to the cheaper coating, i.e., zinc coating, because zinc coating can fail in operating temperatures above 120°C, applicable for 20 of the valves. Thus, the coating should be always standardized to the better-quality coating if the choice is between two types of coating. In another example, let's assume that we have ten, 6″ manual valves in pressure class 300 equal to 50 bar in 22Cr duplex material. Five pieces out of ten are insulated in operating temperatures <100°C. Therefore, the first five valves should be coated with phenolic epoxy, NORSOK coating system 6C. The other five valves have an operating temperature of less than 100°C without any insulation, so they do not require any coating. In such a case, to keep all these identical valves similar in terms of coating implementation, the proposed solution is to paint all ten valves with phenolic epoxy coating system 6C. Again, coating standardization for these identical manual valves results in higher cost, since five of the valves would be coated rather than remaining uncoated. In such a case, when some identical manual valves require coating and some do not, the solution with regard to coating standardization is to coat all of them.

Table 6.3 contains a model of a standardized coating system for manual valves based on the information provided in Table 6.2. Standardization facilitates the coating selection for manual valves.

As discussed earlier in this chapter, the important point is that coating standardization is applicable for valves used topside. Subsea valves are submerged in water, and all of them can be coated with NORSOK system 7, which is two-component epoxy, whether they are in carbon steel, low-alloy steel or stainless steel. Topside actuators are normally in steel material and have operating temperatures of <120°C, so they are coated with either organic or inorganic zinc-rich coating.

TABLE 6.3

Coating System Standardization for Identical Manual Valves in a Topside Offshore Project

Valve Body Material	Conditions	Coating
Carbon steel	All conditions	TSA
Stainless steel 316	All conditions	Phenolic epoxy
22Cr duplex	Operating temperature > 100°C	TSA
	Operating temperature ≤ 100°C	Phenolic epoxy
25Cr duplex	Operating temperature ≤ 20°C	Phenolic epoxy
6MO	All conditions	TSA
Titanium		Not applicable
NAB		Not applicable

6.2.1.3 Valve Color Coding in Firefighting Systems

It is very common to provide color coding for valves and piping components in topside firefighting systems for safety reasons. The selected color is typically red and could be RAL 3020; RAL is the abbreviation of *Reichs-Ausschuß für Lieferbedingungen und Gütesicherung* and provides an identification of the color. Color coding makes it easy for operators and personnel on site to distinguish the piping system, including valves, that supplies water for fire extinguishing purposes. The valve parts that should be coated in red in firefighting systems are handwheels and gearboxes as a minimum. The valve body could also be coated in red. Figure 6.3 illustrates small, 4″ butterfly valves in pressure class 150 equal to 20 bar and titanium material for the body and internals for firefighting seawater service. Handwheels and gearboxes of the valves are color-coded in red to identify the firefighting system. Handwheels and gearboxes of the valves are in stainless steel 316 material.

6.2.1.4 Fire Nuts versus Insulation Boxes on Valves and Flanges

From a valve management point of view, applying paint adds to the cost of painting and increases the total cost of the valve. In addition, applying paint on a valve increases the delivery time of the valve, and time is money. Some valve factories do not have in-house coating capability, in which case the valves are shipped to a coating subcontractor or sub-supplier for coating implementation after the valve pressure test and then returned to the valve factory. As discussed in Chapter 1, piping and valves can be insulated with non-metallic materials or insulation boxes for different reasons, such as increasing the resistance and functionality of the valve in case of fire. In fact, valves and flanges should maintain their integrity in the event of fire by an approach such as fire insulation installation. Insulating piping and valves result in a dangerous type of corrosion for these components, known as corrosion under insulation (CUI). In fact, water or other corrosive compounds can pass through the insulation and get trapped beneath it, causing

FIGURE 6.3 Firefighting butterfly valves with handwheels and gearboxes in red color. (Photograph by author.)

CUI. More information about CUI is provided in Chapter 1. Figure 6.4 illustrates CUI on a piece of pipe. The better solution instead of using an insulation firebox around valves and flanges is to use a fire nut. A fire nut is a kind of fire-resistant nut installed around flange and valve bolts and nuts to protect them from fire (see Figure 6.5). There is no need to install a heavy, bulky and expensive insulation box around flanges and valves when fire nuts are installed. In addition, if fire nuts are installed for valves, it is not necessary to paint them. A fire nut designed to withstand both fire pool and jet fire has additional benefits for the bolts and nuts, such as corrosion and wearing protection. Fire nuts can be easily installed by hand and are normally maintenance free. When fire nuts are installed on the bolts and nuts used to connect valve body pieces, it is important to make sure that there is enough space to install them around the bolt and nut. Figure 6.6 illustrates bolting for ball valve body pieces that are very close to each other, so there is not enough space for fire nut installation.

6.2.1.5 Thermal-Spray Coating Thickness Inspection Challenge for Duplex and Super Duplex Materials

Thermal-spray aluminum (TSA), NORSOK coating system 2A, is applicable for duplex and super duplex piping and valves with operating temperatures above 100°C and 110°C, respectively. The minimum thickness of TSA as per NORSOK

FIGURE 6.4   Corrosion under insulation on a pipe. (Courtesy: Shutterstock.)

FIGURE 6.5   Fire nut around a bolt and nut. (Photograph by author.)

M-501, coating standard, should be 200 μm. Aluminum silicon is a very common sealer applied on TSA coating. TSA has a very porous surface that can be filled in by corrosive fluids or compounds and cause corrosion of the coating, so the porous areas should be filled with a sealant to prevent corrosion. Figure 6.7 illustrates an inspector measuring the TSA coating on the body of a modular valve in duplex material with a paint inspection gauge. The measured thickness is 237 μm, which is acceptable with regard to the required TSA thickness in the NORSOK standard since it is over 200 μm. However, in practice, it has been experienced that measuring the coating thickness on duplex and super duplex with a gauge is not accurate due to the dual ferrite and austenitic properties of the duplex and super duplex materials. The thickness measurement in this case was repeated for

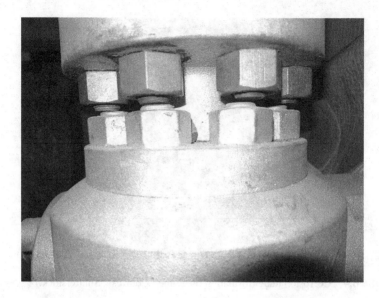

FIGURE 6.6 Lack of space for installation of fire nuts around the bolting of a ball valve. (Photograph by author.)

FIGURE 6.7 TSA coating thickness measurement with a paint inspection gauge on a 22Cr duplex modular valve. (Photograph by author.)

the same valve with different results. However, all the results were acceptable and showed a thickness >200 μm. If the results varied between above and below 200 μm, that would affect the decision of the paint inspector with regard to the sufficiency of the coating thickness, since a coating thickness <200 μm is not acceptable. One way to accurately measure TSA thickness on a duplex valve is to cut a section of the coating on the valve and measure its thickness with a paint inspection gauge.

6.2.1.6   No Coating on Flange Faces

A flange face is where the gasket sits and affects the sealing capability of the flange. The flange face should remain uncoated, since coating the flange can jeopardize the sealability of the flange and cause leakage. There are three types of topside flange faces in general: flat face, raised face and ring-type joint (RTJ) face. For a flat face gasket, the area under the gasket should remain uncoated. If a full-face gasket is selected for a flat-face flange, then no coating is required. On the other hand, the area out of the ring gasket on a flat-face flange should be coated. Figure 6.8 illustrates a flat-face flange with ring and full-face gaskets highlighting the areas that should be coated. It is important to know that it is not recommended to coat the nut-bearing areas and bolt holes based on the present author's experience. More information about the rationale behind keeping nut-bearing areas and bolt holes coating-free is provided in the next section. Figure 6.9 illustrates a raised face flange and highlights the areas that require coating and the areas to remain uncoated. The internal circle of the flange, which is raised and highlighted in the figure, is the place where the gasket sits; it should remain uncoated. The external part of the raised face flange, excluding the bolt holes and nuts, should be coated. Figure 6.10 illustrates an RTJ flange with a metallic gasket and highlights the areas of the flange face that require coating. The metallic gasket sits inside the flange face groove, which should remain uncoated. The external part of the RTJ flange, excluding the bolt holes and nut-bearing areas, requires coating. The internal part of the RTJ flange between the flange middle hole and the RTJ groove should remain uncoated.

A valve supplier used the plastic covers shown in Figure 6.11 to cover the RTJ flanges grooves to prevent them from being coated. Figure 6.12 illustrates a flange

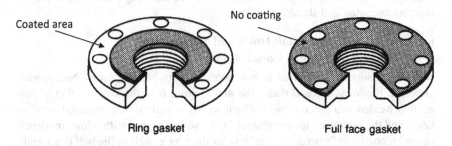

FIGURE 6.8   Ring and full-face gaskets for a flat face flange. (Photograph by author.)

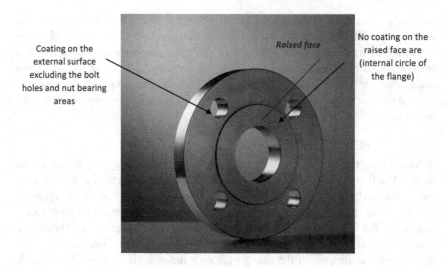

Coating on the external surface excluding the bolt holes and nut bearing areas

Raised face

No coating on the raised face are (internal circle of the flange)

FIGURE 6.9 Raised face flange highlighting the area that requires coating and remains uncoated. (Courtesy: Shutterstock.)

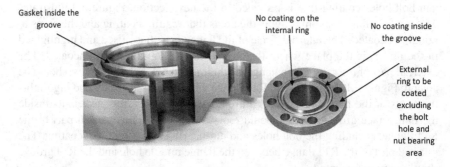

Gasket inside the groove

No coating on the internal ring

No coating inside the groove

External ring to be coated excluding the bolt hole and nut bearing area

FIGURE 6.10 RTJ flange face and metallic gasket highlighting the area that requires coating and remains uncoated. (Photograph by author.)

coated with TSA. The problem is that an area of the flange outside the groove remains uncoated and should be coated.

6.2.1.7   No Coating on Bolt Holes and Nut-Bearing Areas on Flanged Connections

No coating should be applied to bolt holes and nut-bearing areas. Nut-bearing areas are the areas located under the nuts. Figure 6.13 illustrates a flange connection welded to a piece of pipe. The flange is in carbon steel material, uninsulated, and the operating temperature is 70°C, so it is coated with white, inorganic zinc-rich coating. However, the areas under the nuts as well as the bolt holes highlighted in dark red remain uncoated. Why should the bolt holes and nut-bearing

FIGURE 6.11   Plastic covers to cover RTJ flange grooves. (Photograph by author.)

FIGURE 6.12   Uncoated area of a flange coated with TSA that should be coated. (Photograph by author.)

areas remain uncoated? Because coating these areas would affect the tight joint-ing of the bolts and nuts.

The nut-bearing areas should be covered with a layer of primer to prevent corrosion. The main question is how to hide the nut-bearing areas and bolt holes during coating application. Figure 6.14 illustrates bolt plugs, which are placed inside the bolt holes to cover the nut-bearing and bolt hole areas, which should remain uncoated.

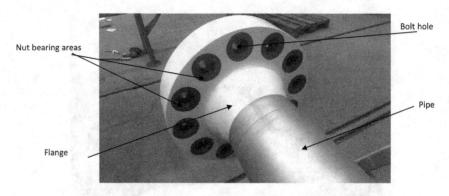

FIGURE 6.13   No coating on nut-bearing areas and bolts. (Photograph by author.)

FIGURE 6.14   Bolt plugs to cover bolt holes and nut-bearing areas. (Photograph by author.)

6.2.1.8   No Coating under the Clamp in Mechanical Joints

Mechanical joints, also called hubs and clamps, can be used in the offshore industry instead of standard flange connections as per the American Society of Mechanical Engineers (ASME) to save weight and space, as illustrated in Figure 6.15.

A mechanical joint, as illustrated in Figure 6.16, includes two hubs, two clamps, four bolts, eight nuts and a seal ring or gasket installed between the two hubs. Like flanges, the hub faces where the seal ring sits should not be coated. The area of the hub located under the clamp is called the shoulder, which is an important section of mechanical joints. The shoulder or hub shoulder is the area that receives the clamp loads and ensures that they are evenly spread around the

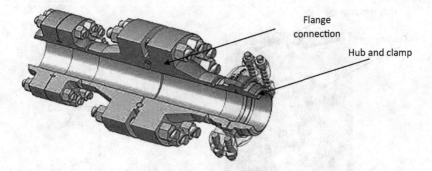

FIGURE 6.15   Comparison between flange and hub and clamp. (Courtesy: SFF/Galperti.)

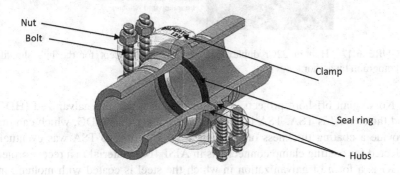

FIGURE 6.16   Mechanical joint. (Courtesy: SFF/Galperti.)

hubs. In fact, the shoulder is very critical in ensuring that the bolt load is transferred to the face of the hub and seal ring. If coating is applied on the shoulder, it can make the assembly of the mechanical joint difficult. Mounting the clamp on a coated hub shoulder can remove the coating in most cases. Even if the clamp mounting did not remove the coating on the hub shoulder, the thermal and pressure transients would eventually break down the coating.

Figure 6.17 illustrates a hub in 22Cr duplex material coated with TSA. As illustrated in the figure, the hub shoulder that stands under the clamp remains uncoated.

6.2.1.9   Clamp Coating Selection

Clamps are parts of mechanical joints mounted on the hubs that grab the hub shoulders. Clamps should have high mechanical strength. AISI 4140, a low-alloy steel with high mechanical strength, is a very common clamp material. AISI 4140 has various alloy elements, such as chromium, molybdenum and manganese. In addition, it has relatively high carbon content, as high as 0.4%, which increases its mechanical strength and hardness. Clamps in AISI 4140 are not corrosion-resistant alloys (CRAs), so they should be coated. Two coating options are typically available according to experiences with mechanical joint suppliers

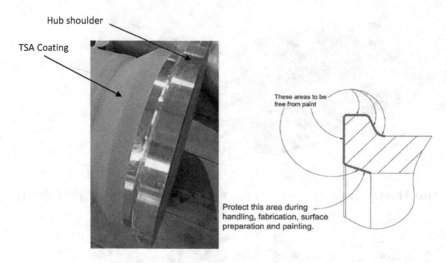

FIGURE 6.17 Hub in 22Cr duplex coated with TSA except for the hub shoulder. (Photograph by author.)

in Norwegian offshore projects. The first option is hot-dip galvanized (HDG), and the second is TSA. TSA is stronger and thicker than HDG, which can only provide a coating thickness of between 40 and 90 μm, so TSA was eventually selected for coating clamp connections in AISI 4140 material in a recent project. HDG is a form of galvanization in which the steel is coated with molten zinc. Figure 6.18 illustrates a clamp connection in AISI 4140 coated with TSA.

6.2.1.10 Minimizing the Usage of HDG Bolting

Low-alloy HDG bolting is widely used for valves, flanges, hubs and clamps in the oil and gas industry, particularly topside offshore. Many low-alloy steel HDG bolts and nuts are used for industrial valves in exotic materials such as duplex, super duplex and 6MO. However, if the HDG is removed from the bolts and nuts, the underlying, non-corrosion-resistant alloy (low-alloy steel) will be corroded after some years of operation. If such bolting is used for valves in exotic materials such as duplex and super duplex, then the valve will be useless after the corrosion of the bolting and will have to be thrown out. Merely screwing or unscrewing the bolts inside the nuts causes metal-to-metal contact and friction between the threaded areas of the bolts and nuts, and the HDG can be removed by these actions even in a short period of time. Figure 6.19 illustrates the removal of HDG from the low-alloy steel nut material installed on a mechanical joint.

The recommendation of the present author, based on industrial experiences, is to limit the usage of low-alloy steel bolting with HDG coating for carbon steel flanges and valves. A 25Cr super duplex bolts and nuts are proposed for use with duplex and super duplex valves. Inconel 625 bolting is an alternative to super duplex bolting where super duplex bolting cannot be used. As an example, if the operating temperature of the valve is over 110°C, super duplex bolting should be

AISI 4140 clamp
with thermal
spray aluminium
coating

FIGURE 6.18 AISI 4140 clamp coated with TSA. (Photograph by author.)

Removal of hot dip
galvanized from the nut

FIGURE 6.19 Removal of HDG from a low-alloy steel nut in a mechanical joint. (Photograph by author.)

upgraded to Inconel 625 bolting, as the super duplex material is at high risk of pitting and CSCC at operating temperatures above 110°C in the offshore environment. More information about pitting and CSCC is provided in Chapter 1. Inconel 625 bolting can be used for valves and flanges in 6MO material in addition to duplex and super duplex valves and flanges in special cases of high temperatures above 110°C. It is important to know that there is no need to apply any coating on corrosion-resistant bolting like super duplex and Inconel 625, unlike low-alloy steel bolts, which are coated with HDG. Figure 6.20 illustrates the 22Cr duplex body of a ball valve with Inconel 625 bolts and nuts instead of super duplex bolting because the operating temperature of the valve is above 110°C. The grade, F467, is engraved on the nut, meaning that the nut is in a nickel alloy such as Inconel 625. F468 is engraved on the bolt end, meaning that the bolt is

Inconel 625
bolt

Inconel 625 nut

FIGURE 6.20   Inconel 625 bolts and nuts for ball valves in 22Cr duplex body material in operating temperatures higher than 110°C. (Photograph by author.)

also in a nickel alloy such as Inconel 625. These and other grades for different materials are defined in the American Society for Tests and Materials (ASTM) standard. Low-alloy steel bolts with zinc-nickel plate coating are very common for subsea valves. The usage of CRAs such as super duplex and Inconel 625 for bolting is not common for subsea valves, because the bolts and nuts for subsea valves are subject to cathodic protection against external corrosion, which can be caused by seawater. Due to the implementation of cathodic protection to prevent external corrosion attack on subsea bolting, there is no need in general to use corrosion-resistant bolts.

6.2.1.11   Coating the Valve after the Pressure Test

The pressure test is a part of the factory acceptance test (FAT) that is performed by the valve supplier on manufactured valves to evaluate the structural integrity and sealability of the valves. The pressure test is mainly divided into two categories: the first is the shell or body test to evaluate valve integrity, and the second is the seat test to evaluate valve sealability. It is very common to apply these pressure tests, especially the body test, before applying coating on the valve. If the valve body is coated during the pressure test, then the coating can hide possible leakage from the valve body. It is important to know that the valve body is pressurized internally with 1.5 times of the valve design pressure during the body pressure test, and the valve body and bonnet are monitored for any possible leakage. No leakage from the body is allowed during the body or shell pressure test. Figure 6.21 illustrates a very large (38″) ball valve with a white electrical actuator installed on top of it during the FAT. The body of the valve is in carbon steel, and it is not coated during the test, enabling the inspector to

FIGURE 6.21 A 38″ ball valve with electrical actuator during a pressure test. (Photograph by author.)

FIGURE 6.22 A 38″ ball valve with electrical actuator after pressure test and coating. (Courtesy: Flow Control Technology (FCT).)

see any possible leakage from the valve during the test. The electrical actuator mounted on top of the valve is in steel coated with inorganic zinc coating. Figure 6.22 illustrates the same 38″ ball valve with its electrical actuator after the pressure test and coating application. The valve body is in carbon steel and the operating temperature is <120°C, so it is coated with inorganic zinc coating in white color.

6.2.1.12 Insulation Boxes

Insulation can be installed like a box around certain valves. Figure 6.23 illustrates a valve installed inside an insulation box. The whole valve assembly, except for the handwheel and gearbox, is installed inside the insulation box. The valve handwheel is outside the insulation box because it is the only means of operating the valve. The gearbox is also located outside the box because it shows the position of the valve (open or closed).

Phenolic epoxy is a very common type of coating on valves located under insulation, as discussed in Chapter 4. Alternatively, TSA coating could be selected for valves under insulation for operating temperatures above 150°C where phenolic epoxy cannot be used. The question is, how can an operator or inspector access a valve installed under insulation for inspection or maintenance purposes? The answer is that the insulation box should have a window or hatch, or the insulation should be removable to provide access to the valve for coating inspection or any other purposes.

6.2.1.13 Coating the Valves' Top Flange

A valve's top flange or mounting flange is a flange on the top of a valve that allows the attachment of the valve to the valve actuator or gearbox. The valve top flange can also act as a support for connected components such as a gearbox or actuator according to the API 6DX standard for actuator sizing and mounting kits for pipeline valves. The mounting flange is typically enclosed with the connected component, so it is not exposed to the corrosive environment in general. However, the gearbox or actuator could be disassembled from the valve for maintenance and repair. In that case, the valve top flange would be exposed to the corrosive off-shore environment. A practice experienced in some offshore projects is to apply at least one layer of epoxy primer to the uncoated areas of the top flange in carbon steel material for corrosion protection. Figure 6.24 illustrates the top flange of a ball valve in carbon steel material. The ball valve, including the outer area of the

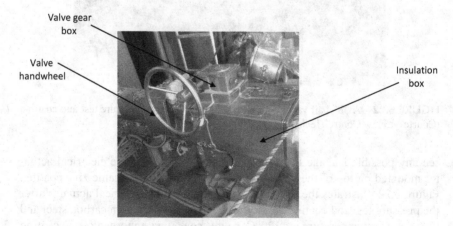

FIGURE 6.23   Insulation box around a ball valve. (Photograph by author.)

TSA coating   One-layer epoxy
                                              primer

                                         Valve top flange

FIGURE 6.24   Ball valve top flange coated with TSA and one-layer epoxy primer. (Photograph by author.)

top flange, is coated with TSA. The inner area of the top flange is coated with one-layer epoxy primer in red color for corrosion protection.

6.2.1.14   Using a Stainless Steel 316 Gearbox

Cast iron or carbon steel valve gearboxes coated with zinc-rich coating were once common in the Norwegian offshore industry. Cast iron is a type of carbon iron alloy with a relatively higher carbon content (higher than 2%) compared to carbon steel. Figure 6.25 illustrates a steel gearbox coated with zinc-rich primer in an offshore environment after some years of operation. Some of the coating has worn off and has become cracked in places, and the gearbox is now corroded. The alternative solution, which provides better corrosion mitigation, is to select the gearbox material in austenitic stainless steel 316, in which case there is no need to apply coating on the gearbox. Figure 6.26 illustrates a gearbox in stainless steel 316 on a ball valve.

6.2.1.15   Lack of TSA Coating Adhesion

The failure of TSA coating to adhere to a steel surface was experienced in one offshore project involving about 20 small gate valves; Figure 6.27 illustrates one of them. The reason for the lack of adhesion was attributed to poor surface preparation. Thus, surface preparation is a very important step before coating application. Many coatings fail at early stages solely because of poor surface preparation. Surface preparation is explained in more detail in Chapter 2.

6.2.1.16   Coating to Prevent Cross-Contamination

Stainless steel materials are at risk of cross-contamination during manufacturing, transportation and storage at the valve factory. There are several ways for stainless steel to be contaminated in the factory, such as through contact with carbon steel. As an example, in the manufacturing process, several different tasks such as welding and grinding can generate dust from carbon steel that can be transferred to other materials such as stainless steel and cause cross-contamination.

FIGURE 6.25 Corroded cast iron gearbox coated with zinc-rich primer in an offshore environment. (Photograph by author.)

Stainless steel is known as a corrosion-resistant alloy that can provide better corrosion resistance compared to steels such as carbon steel in corrosive environments. The main question is, how does cross-contamination affect stainless steel? Stainless steel is corrosion-resistant because it contains a chromium level of at minimum 11%–12%. The chromium can form a protective layer of Cr_2O_3 in reaction with the oxygen in the atmosphere. The protective layer of chromium oxide is the main factor that accounts for the corrosion resistance of stainless steel. Cross-contamination of stainless steel damages this protective layer, such that the stainless steel is no longer corrosion-resistant. Thus, it is important to keep stainless steel materials away from carbon steel during fabrication, assembly, transportation and storage. Figure 6.28 illustrates a 22Cr duplex body valve coated with Teflon powder. Teflon powder is used on the valve body to prevent the cross-contamination of duplex valves in case they come in contact with carbon

FIGURE 6.26 Stainless steel 316 gearbox installed on a ball valve. (Photograph by author.)

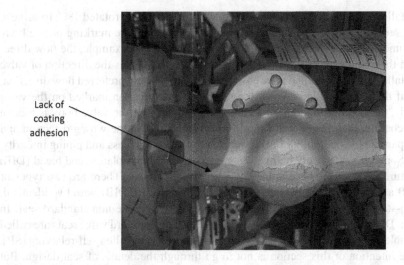

FIGURE 6.27 Lack of TSA coating adhesion to the substrate in small gate valves. (Photograph by author.)

steel materials. It is essential to note that the Teflon powder coating used for this application is not a kind of paint.

6.2.1.17 Coating and Poor Marking on Valves

Some important signs should be marked on valve bodies, such as flow direction. Marking the flow direction is applicable to unidirectional valves to show the correct direction for valve installation. Unidirectional valves are those that should be installed in one direction or one position. Bidirectional valves, conversely, can be

FIGURE 6.28 Duplex valve coated with Teflon powder to prevent cross-contamination. (Photograph by author.)

installed in two directions, meaning that the valve can be rotated 180° to achieve the second position of installation. It is important that the marking on the body of unidirectional valves is not hidden by coating. As an example, the flow direction that can be marked on check or globe valves indicates the direction of valve installation. As explained previously, check valves have a preferred flow direction, so if they are installed against the required flow direction marked on the valve body, they do not function properly. If the marking on the valve body is hidden by coating, then the operator might install the valve in the wrong direction and jeopardize the operability of the valve as well as the process and piping integrity.

Some valves, mainly ball valves, may have a double isolation and bleed (DIB) feature. DIB valves have at least one non-standard seat. There are two types of DIB according to API 6D, the pipeline valves standard: DIB1 with two identical, non-standard seats; and DIB2 with one standard and one non-standard seat. In fact, DIB1 is bidirectional but DIB2 is not. The non-standard valve seats are called double piston effect (DPE), while the standard seats are called self-relieving (SR). The intention of this section is not to go through the details of seat design. But in general, DPE seats can provide tighter sealing compared to SR seats. From a maintenance point of view, if the piping connected to the valve is under maintenance, a DPE seat can provide greater safety for the operator doing the maintenance compared to an SR seat. DIB2 valves with one DPE and one SR seat are not bidirectional, meaning that they have a correct direction of installation in only one position. Thus, the DPE and SR seats should be marked on the body of DIB2 valves so it is possible to distinguish how the valves should be installed on the line. The reason why a unidirectional DIB2 valve is preferred over a bidirectional DIB1 could be related to lower cost and simpler design. However, it should be noted that a DIB1 valve should be selected in cases where the operation can take place on both sides of the valve. It is best practice to install the valve such that the

FIGURE 6.29   DIB1 actuated ball valve with one SR and one DPE seat and highlighting the maintenance side on DPE seat direction. (Courtesy: Shutterstock.)

FIGURE 6.30   Engraved DP on the flange body of a ball valve; inappropriate marking. (Photograph by author.)

DPE seat is toward the side where maintenance takes place. Figure 6.29 illustrates an actuated DIB2 ball valve with one DPE and one SR seat. The maintenance should take place on the left side of the valve where the DPE seat is located. Figure 6.30 illustrates a section of a DIB2 valve with one SR and one DPE seat. DP, as an indication of DPE, is marked on the right flange body of the valve with a red marker and a small engraving. The marking illustrated in the figure is not appropriate, as it can be hidden by the valve coating. In such a case, without correct identification of the seats, it is possible to install the valve in the wrong direction, which will put the operator's health and safety in danger and jeopardize the operability of the system.

The correct method of marking the body of a valve is provided in the API 6D standard. A name or tag plate in stainless steel 316 material should be installed on the body of the valve after coating, and the tag plate should be affixed to the valve body with four rivets in stainless steel 316. Using steel or carbon steel for the tag plate and rivets is not recommended, because they can be corroded easily. Figure 6.31 illustrates the proposed tag plate shape and format specified in API 6D. The left side of the tag plate shows one arrow indicating an SR seat, and the right side shows double arrows indicating the DPE seat. There are four holes on the four corners of the tag plate for rivet installation.

6.2.1.18   Coating Close to Welded Areas

Welding is a very common method of joining pipes together in the fabrication yard. The most common type of welding is the buttweld connection. The buttwelding ends are prepared by beveling each end of the pipe or valve to match the similar beveled pipe to which it will be attached. Valves could be welded to the piping; in such a case, the valve is supplied with a pup piece at both ends. The coating of the valve and pipe is typically performed by the valve and piping supplier or a coating subcontractor company before sending the components to the fabrication yard for fabrication, welding and assembly. Figure 6.32 illustrates a non-slam axial nozzle check valve that will be installed on a pipeline after a compressor in a compressor station unit. A compressor is a piece of mechanical equipment used to increase the pressure of gas in piping services. The valve in the figure has two pup pieces, one at each end, which are buttwelded to the connected piping in the yard. The pup pieces provide some distance between the valve internals and the

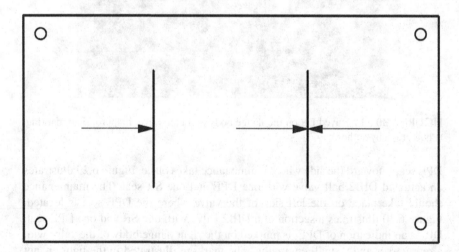

FIGURE 6.31   Tag plate based on API 6D to indicate the SR and DPE seats on a DIB2 valve. (Courtesy: American Petroleum Institute.)

Pup piece

Pup piece

End of pup piece no coating

End of pup piece no coating

FIGURE 6.32 Axial nozzle check valve welded to a pipe from both ends. (Photograph by author.)

points where the valve is welded to the piping in the yard. Welding the valve to the piping produces a considerable amount of heat, which can damage the valve seals. The valve seals are typically in soft or non-metallic materials that can be melted because of exposure to high temperature. In addition, the produced heat can cause thermal expansion and reduce the volume of the valve internals, such that the valve internals could lose their sealability. For all these reasons, pup pieces are integrated into the valve design to keep the valve internals, including seals, away from the heat produced during the welding and to prevent possible leakage from the valve during operation. As you can see in the figure, the whole valve including the pup pieces is coated in the valve manufacturer factory. A possible coating problem is that the coating could be damaged and removed in those areas close to the ends of the valve pup pieces after they are welded to the connected piping in the yard. The recommendation is to not apply coating on the pup pieces in the area <50 mm from the end parts welded to the piping. To protect the uncoated areas of the valve pup piece ends from corrosion, it is recommended to protect them with a primer.

6.2.2 Topside Actuator Coating Case Studies

6.2.2.1 Actuator Color-Coding

Actuators for topside can be color-coded to indicate their safety functions; fail-safe open actuators can be coated with green color (RAL 6002). Fail-safe closed actuators can be colored in red (RAL 3000). Fail-as-is actuators, including electrical motors or double-acting actuators, can be colored in gray white (RAL 9002). Figure 6.33 illustrates a 38″ ball valve in carbon steel material and an operating temperature of 70°C, so inorganic zinc coating is applied on the valve. The valve is operated by an electrical actuator installed on the top of the valve. The electrical actuator is in steel and coated with inorganic zinc coating. The electrical

FIGURE 6.33 A 38″ ball valve in carbon steel material with an electrical actuator; both valve and actuator are coated with inorganic zinc coating. (Photograph by author.)

FIGURE 6.34 Ball valve in carbon steel with pneumatic actuator coated red with inorganic zinc epoxy. (Photograph by author.)

actuator is fail-as-is, so the color of the coating is white gray. Figure 6.34 illustrates another carbon steel valve coated with inorganic zinc epoxy; the valve has a pneumatic actuator. The actuator is coated with inorganic zinc epoxy, like the valve, but in red because it has a fail-safe closed function.

6.2.2.2 Coating the Area under the Actuator End Stopper

End stoppers are installed on both sides of pneumatic and hydraulic actuators. Figure 6.35 illustrates an end stopper at the end of a pneumatic actuator. The main function of an end stopper is to make sure that the valve is completely placed in

FIGURE 6.35 End stopper at the end of a pneumatic actuator. (Photograph by author.)

FIGURE 6.36 Lack of coating under the end stopper of an actuator. (Photograph by author.)

open or closed position. As an example, a ball valve requires a 90° rotation of the ball to move between open and closed positions. Actuators are designed and tested to ensure that they can provide the required ball movement and rotation between open and closed. Let's suppose, however, that a pneumatic actuator rotates the ball only 89° instead of 90° to open position. A human operator could use the end stopper to adjust the valve ball to fully open position. In fact, an end stopper can be rotated by the operator to move the ball 1° to keep the valve in a fully open position. It has been experienced in offshore projects that the area under the end stopper remains uncoated, as illustrated in Figure 6.36. However, lack of coating under the actuator end stopper causes corrosion of the actuator during operation. Thus, the recommendation is to remove the end stopper from the actuator during coating application to make sure that the area under the end stopper is coated completely.

6.2.2.3 Handwheel Coating for Valves and Electrical Actuators

Handwheels on valves and electrical actuators should be in stainless steel 316. Carbon steel handwheels can be corroded easily in the offshore environment.

Figure 6.37 illustrates rust and general corrosion on a valve handwheel in carbon steel material in the offshore environment.

Figure 6.38 illustrates an electrical actuator. Its handwheel had been supplied with HDG carbon steel initially from the actuator supplier. HDG carbon steel cannot provide sufficient corrosion resistance in the corrosive offshore environment, since when the zinc coating (HDG) is removed from the handwheel, the carbon steel can be corroded easily. Thus, TSA should be sprayed on the handwheel in such a case to mitigate external corrosion. The

FIGURE 6.37 Rust and general corrosion on a carbon steel valve handwheel. (Courtesy: Shutterstock.)

FIGURE 6.38 Electrical actuator with a handwheel in carbon steel plus HDG coated with TSA to mitigate external offshore corrosion. (Photograph by author.)

FIGURE 6.39 Pneumatic actuated ball valve on a test bench. (Photograph by author.)

ideal solution is to supply electrical actuators in stainless steel 316 from the beginning.

6.2.2.4 Actuator Disassembly from the Valve during Coating

Actuated valves are typically tested by the valve manufacturer at a factory when the actuator is mounted on the valve. In general, actuated valves are tested in the valve factory before sending the valve for coating. Figure 6.39 illustrates the testing of a pneumatic scotch and yoke actuated valve on a test bench. The actuator had been pressure tested individually and coated before being mounted on the valve. Some actuator tests, such as pressure tests, are done individually before coating the actuator. The actuator is pressure tested, coated and then mounted on the valve. Valve tests, as well as some of the actuator tests, such as function tests, are performed with the actuator mounted on the valve. Actuators function tests are those in which the actuator is used to operate (open and close) the valve in order to make sure the actuator is functioning properly.

The actuated valve is sent for coating after the tests, and the actuator should be disassembled from the valve prior to the valve coating. Disassembling the actuator from the valve before valve coating could jeopardize the results and validity of the function test that had been performed while the valve and actuator were tested together. One solution is to postpone the actuated valve function tests until after the valve is coated.

6.2.2.5 Passive-Fire Protection around a Double-Acting Hydraulic Actuator

A very large ball valve installed on an oil export pipeline in 38″ and a pressure class of 1,500 equal to 250 bar had a hydraulic actuator for automation, as shown in Figure 6.40. Initially, a hydraulic scotch and yoke spring return or single-acting actuator were selected for this large, high-pressure class valve. The spring-return hydraulic actuator had a fail-safe closed function that was achieved by spring force. The force provided by the spring of the actuator was

FIGURE 6.40 A 38″ CL1500 ball valve on an oil export pipeline with a double-acting actuator. (Photograph by author.)

not sufficient to close such a large valve, however, so a decision was made to upgrade to a double-acting actuator to provide greater force for the valve operation. As explained in Chapter 5, a double-acting actuator is opened and closed by hydraulic oil force. One disadvantage of a double-acting actuator compared to a single-acting or spring-return actuator is related to operability in the event of fire. A spring-return actuator returns to fail-safe closed position and closes the valve in case of fire. However, a double-acting actuator is springless, so the hydraulic oil is burnt in the event of fire after a short period of time, and the actuator cannot move to return the valve to closed position. The solution is to use passive-fire protection (PFP) around a hydraulic actuator to ensure its functionality for a specific period after a fire begins. The PFP could be either NORSOK coating system 5 or an enclosed container such as a box around the actuator. The decision in this case was to apply a PFP box around the double-acting actuator, as illustrated in Figure 6.41. Since PFP was applied around the actuator, it was not coated with NORSOK system 5, PFP coating. Alternatively, inorganic zinc-rich coating system 1A was applied to the actuator.

6.3 COATING APPLICATIONS FOR SUBSEA VALVES AND ACTUATORS

6.3.1 Subsea Valves

6.3.1.1 Subsea Valve Coating

Figure 6.42 illustrates small subsea ball valves in sizes <1″ used for hydraulic and chemical lines in a subsea distribution unit (SDU). An SDU is a kind of manifold that includes a steel structure, piping and mainly small valves to transport

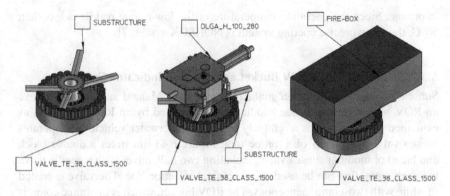

FIGURE 6.41 PFP box installed around the double-acting actuator of a 38″ ball valve in CL1500 for an oil export pipeline. (Photograph by author.)

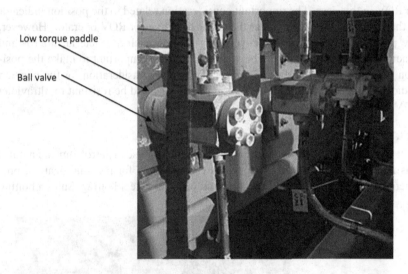

FIGURE 6.42 Small ball valves with low-torque paddles in an SDU coated with two-component epoxy. (Photograph by author.)

and direct chemicals and hydraulic fluid to the well-head and other manifolds. Hydraulic fluid is used for the operation of the actuators. Chemicals are injected into the process lines to prevent operational problems like wax and scale formation, etc. All the valves are in 25Cr super duplex and are coated with NORSOK system 7, two-component epoxy suitable for subsea use. The coating of the valves is all in white color. The ball valves have an operating temperature above 50°C, so the more precise coating system according to the NORSOK M-501 standard is coating system 7C. All of these small valves have a low-torque paddle installed on top so that they can be operated by a remote-operated vehicle (ROV). The low-torque paddle is coated with two-component epoxy, just like the valves, but in orange. Since the operating temperature of the low-torque paddle is less than 50°C, the more precise coating system is NORSOK system 7B.

6.3.1.2 Subsea Valve ROV Bucket and Position Indicator Coating

Subsea valves could be either manual or actuated. Manual subsea valves have an ROV bucket or low-torque paddle that is operated by an ROV. An ROV, as explained in Chapter 5, is a remotely operated underwater vehicle that operates subsea valves by means of a torque tool. Figure 6.43 illustrates a double block and bleed or modular subsea valve, including two ball valves with a needle valve between them, that can be used on a chemical injection line. The valve is coated in white with two-component epoxy. The ROV bucket, which is in orange color, is also coated with two-component epoxy. There is a position indicator close to the ROV bucket that is coated with two-component epoxy in black color. The position indicator shows the position of the valve (open or closed) to the operator, which is an ROV in this case. The important consideration related to the position indicator is that it should be visible under the water to a diver or ROV operator. However, the position indicator can be covered by fouling such as bacteria, micro- and macro-structures and may not be visible. Thus, it is important to make the position indicator in anti-fouling materials. The other consideration for the position indicator in terms of material selection is that it should be resistant to ultraviolet (UV) light.

6.3.1.3 Subsea Valve Bolting

API 20 E is titled, "Alloy and carbon steel bolting for use in petroleum and natural gas industries." This standard specifies requirements for the qualification, production and documentation of carbon and low-alloy steel bolting. Subsea bolting

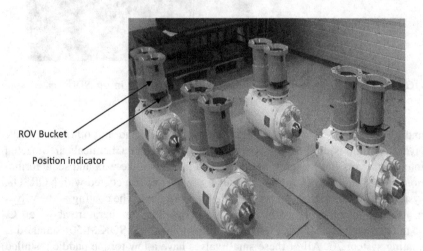

ROV Bucket

Position indicator

FIGURE 6.43 ROV-operated modular valves coated with two-component epoxy. (Photograph by author.)

components (bolts and nuts) are subject to cathodic protection to prevent external corrosion due to contact with seawater. Therefore, it is not common to use exotic bolting materials such as super duplex or Inconel 625 for subsea valves. API 20 E defines three different bolting specification levels (BSL): BSL1, BSL2 and BSL3. Each BSL defines the level of technical quality and the qualification level of the bolting. Higher BSL (BSL3) reflects higher-quality bolting. In general, BSL1 bolting should satisfy the requirements given in the related ASTM standard, such as ASTM A193, A194 or A320. BSL2 requires additional testing compared to BSL1 and has more restricted requirements. BSL3 is the most restricted level with additional volumetric testing that is not required for BSL1 and BSL2. BSL1 is not proposed for subsea valve application. Zinc nickel plating is a very common type of low-alloy steel bolting coating protection in the subsea oil and gas industry. As explained earlier in this chapter, HDG (molten zinc coting) is a common type of low-alloy steel bolt coating in the topside offshore oil and gas industry. Figure 6.44 illustrates bolt and nuts for a subsea gate valve

FIGURE 6.44 Low-alloy steel bolts and nuts for a subsea gate valve with zinc-nickel electroplating. (Photograph by author.)

in low-alloy steel electroplated with zinc nickel. The electroplating process is explained in Chapter 2.

API 20 F is titled, "Corrosion resistant bolting for use in the petroleum and natural gas industries." This standard specifies requirements for the qualification, production and documentation of corrosion-resistant bolting used in the oil and gas industry. This standard is applicable for equipment and components that are covered by API standards. API 20 F introduces just two BSLs: BSL2 and BSL3. In fact, BSL1 is not applicable for corrosion-resistant bolting such as ASTM A453 Gr. 660 and API 6A718. ASTM A453 Gr. 660 is a nickel alloy bolting that contains 25% nickel and 15% chromium. API 6A718 covers Inconel 718 bolting, a hard nickel and chromium alloy bolting with high mechanical strength. Using this bolting material compared to Inconel 625 bolting, which has less mechanical strength, could lead to using fewer bolts for the body/bonnet or body pieces of a valve connection, which could save some area on the body/bonnet of the valves. As noted above, API 20F and exotic bolting are not common in the subsea industry. The bolting grades mentioned above as per API 20F are more common for topside valves.

6.3.2 SUBSEA ACTUATORS

6.3.2.1 Subsea Actuator Coating

Figure 6.45 illustrates a 7 1/16″ through conduit gate valve installed on the branch line of a subsea manifold. The gate valve is automated with a linear hydraulic actuator installed vertically on top of the valve. Both valve and actuator are coated with two-component epoxy in white color for external corrosion protection in the subsea zone. The operating temperature of the valve is higher than 50°C, so the applicable coating system is NORSOK system 7C, but the operating temperature of the actuator is <50°C, so the applicable coating system is NORSOK system 7B.

6.3.2.2 ROV Override Coating

Subsea actuators are equipped with ROV override. The main purpose of ROV override is to operate the valve independently from the actuator. It means that if the actuator were disassembled from the valve for maintenance, an operator could operate the valve by means of ROV override. Figure 6.46 illustrates a subsea actuator coated in white with two-component epoxy. The ROV override installed on the actuator is coated with two-component epoxy in orange color.

6.4 MORE CASES AND FIGURES

Figures 6.47–6.57 provide more pictures related to the coating of piping, valves and actuators. Figure 6.47 shows the RTJ flange end of a valve. The valve is in 22Cr duplex material with an operating temperature higher than 100°C, so the valve is coated with TSA. The RTJ groove and face of the flange remain uncoated as they are illustrated.

Linear actuator

Subsea gate valve

FIGURE 6.45 Subsea gate valve with linear actuator, both coated with two-component epoxy. (Photograph by author.)

FIGURE 6.46 Subsea actuator and ROV override coated with two-component epoxy. (Photograph by author.)

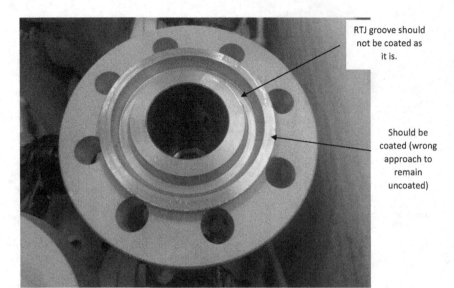

RTJ groove should
not be coated as
it is.

Should be
coated (wrong
approach to
remain
uncoated)

FIGURE 6.47   RTJ flange end of a valve. (Photograph by author.)

FIGURE 6.48   A 4″ globe valve during a pressure test; the valve should remain uncoated
during the pressure test as shown in the figure. (Photograph by author.)

Figure 6.48 shows a 4″ globe valve during a pressure test. The valve is not
coated during the pressure test, as coating can hide possible leakage from the
valve body during the pressure test. Figure 6.49 illustrates a small wedge gate
valve with a 22Cr duplex body in an operating temperature above 100°C, coated
with TSA. Figure 6.50 illustrates a small wedge gate valve in carbon steel and

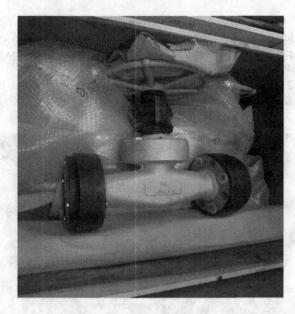

FIGURE 6.49 Small gate valve in 22Cr duplex coated with TSA. (Photograph by author.)

FIGURE 6.50 A small wedge gate valve in carbon steel and an operating temperature <120°C coated with inorganic zinc-rich coating. (Photograph by author.)

an operating temperature <120°C coated with inorganic zinc-rich coating. Figure 6.51 illustrates an inspector measuring the thickness of TSA coating on a small wedge gate valve with a gauge. The gauge shows the thickness of the coating equal to 249 μm.

Figure 6.52 shows an axial nozzle check valve in 22Cr duplex coated with TSA because the operating temperature of the valve is above 100°C. Figure 6.53

FIGURE 6.51 Inspector measuring the thickness of TSA coating on a small wedge gate valve. (Photograph by author.)

FIGURE 6.52 Axial nozzle check valve in 22Cr duplex coated with TSA. (Photograph by author.)

FIGURE 6.53   A worker prepares a dual-plate check valve for sandblasting. (Photograph by author.)

Clamp
shoulders
remain
uncoated

FIGURE 6.54   Hub-ended axial nozzle check valve coated with TSA except for the clamp shoulders. (Photograph by author.)

illustrates a worker preparing a dual-plate check valve for sandblasting before coating implementation. Both ends of the dual-plate check valve, which are machined precisely, are covered with plastic to prevent them from being damaged by the sand abrasives during sandblasting. Figure 6.54 illustrates an axial nozzle check valve with hub ends coated with TSA, except for the clamp shoulders or under clamp areas.

Figure 6.55 illustrates a ball valve with a pneumatic actuator during a test. The valve is uncoated during the test, and the actuator is coated with zinc-rich coating in red color since it has a fail-safe closed function. Figure 6.56 illustrates ball valves in carbon steel coated with inorganic zinc-rich coating and gearboxes in stainless steel 316 material. Figure 6.57 shows butterfly valves in titanium material for fire water service; valves' handwheels are colored red to identify their function in fire water service.

FIGURE 6.55   An actuated valve, the valve remaining uncoated during the pressure test. (Photograph by author.)

FIGURE 6.56   Carbon steel ball valves coated with inorganic zinc coating, with stainless steel 316 gearboxes. (Photograph by author.)

FIGURE 6.57   Titanium butterfly valves for firewater service and red color code gearbox for identification. (Photograph by author.)

6.5   CONCLUSION

This chapter is written according to the industrial experiences and contains many different cases related to the coating of piping, valves and actuators for the off-shore industry. This chapter is illustrated with figures based on the industrial experiences.

6.6   QUESTIONS AND ANSWERS

1. Which sentence is correct about the application of coating on valves and actuators in the offshore oil and gas industry?
   A. Applying coating increases the total cost of valves and actuators, considering both the cost of expenditure (initial cost) and the operation cost.
   B. The cost of applying coating on NAB is higher than on duplex and super duplex.
   C. Coating standardization can increase the total coating cost.
   D. Subsea valves for firefighting systems must be coated in red color for color-coding purposes.
   Answer: Option A is not correct, because coating increases the initial cost but reduces the operation and maintenance cost that can be imposed during the project because of external corrosion damage. Option B is not correct, since NAB requires no coating while duplex and super duplex may require TSA or phenolic epoxy depending on the operating conditions. In fact, then, the cost of coating for NAB is zero. Option C is correct; coating standardization can increase the cost of coating. As an

example, carbon steel valves require inorganic zinc coating for operating temperatures less than and equal to 120°C. However, a stronger and more expensive coating such as TSA is applied on carbon steel valves in operating temperatures above 120°C. It is not possible to standardize the coating of valves to inorganic zinc coating since this type of coating is not suitable for valves at operating temperatures above 120°C. The coating that should be selected as a standard solution in this case is TSA, which increases the cost of coating of the valves in total. Option D is not correct, because there is no firefighting system subsea. Topside handwheels and gearboxes for valves used in firefighting systems are color-coded with red coating. Thus, option C is the correct answer.

2. Which piping and valve areas should remain uncoated based on the information provided in this chapter?
   A. Flange faces, the area under the clamp on the hub shoulder
   B. Under nuts or nut-bearing area on flanges
   C. Bolt holes in flange connections
   D. All of the above

   Answer: Bolt holes and nut-bearing areas on flanges, as well as flange faces, should remain uncoated. Coating these areas could jeopardize the sealing of the flange joints. In addition, coating the areas under the clamp on the hub shoulder is useless, since if the areas under the clamp are coated, the coating will be removed by the clamp force eventually. Thus, option D, which includes all the options, is the correct answer.

   Answer: Option A is not correct because phenolic epoxy is the suitable coating system under insulation for operating temperatures of at maximum 150°C. However, in this case, the operating temperature is 160°C, so the correct coating system is TSA. Option B is not correct either, because two-component epoxy is suitable for subsea valves and actuators. The correct coating system for a steel topside actuator is inorganic zinc epoxy, and it can be red in color, since the actuator is fail-safe closed. Option C is not correct because inorganic zinc-rich coating is a suitable coating for operating temperatures of at maximum 120°C for uninsulated carbon steel material components such as piping and valves. TSA is the proposed coating system for uninsulated carbon steel materials in operating temperatures higher than 120°C. This means that the standardization of all piping and valves in this case to inorganic zinc-rich coating would lead to the coating and material failure of any piping and valves operating between 120°C and 150°C. Thus, the correct standardization coating in this case is TSA. Option D is the correct answer; 22Cr duplex at 170°C, whether insulated or not, should be coated with TSA. The hub face and shoulder areas, as illustrated in Figure 6.17, should remain uncoated.

4. Fill in the gaps with the correct words.
   A 10″ ball valve with a super duplex body is supplied with low-alloy steel bolting coated with HDG. The better material selection for the

bolting would be _____, assuming that the operating temperature of the valve is 90°C. In another case, a carbon steel gate valve is coated with inorganic zinc coating because the operating temperature of the valve is less than _____. The top flange of the gate valve should be coated with _____. The inner part of the top flange can be coated with one layer of _____ primer.
A.   Super duplex, 120°C, inorganic zinc coating, epoxy
B.   Inconel 625, 150°C, TSA, TSA
C.   Super duplex, 130°C, phenolic epoxy, phenolic epoxy
D.   Inconel 625, 140°C, epoxy coating, epoxy

  Answer: HDG low-alloy steel bolting is not suitable for super duplex body valves. The HDG coating can be removed from bolts and nuts after some period of operation, at which point the bolting would be at high risk of corrosion. In the event of bolting failure due to corrosion, the expensive valve in the exotic material of super duplex would useless and should be thrown away. Thus, super duplex bolting is proposed for super duplex valve bodies, considering the operating temperature of 90°C. However, super duplex bolting has a temperature limitation of at maximum 110°C, meaning that Inconel 625 bolting should be selected for operating temperatures higher than 110°C. In the second case, a carbon steel valve is coated with inorganic zinc coating because the operating temperature of the valve is <120°C. Inorganic zinc coating can be used on carbon steel materials for operating temperatures of at maximum 120°C. The top flange can be coated with the same coating used for the valve body, which is inorganic zinc coating on the outer part. The inner part of the top flange is coated with epoxy. Thus, option A is the correct answer.

5. Identify the incorrect statements below.
  A.   A gearbox in steel or cast iron coated with inorganic zinc is subject to corrosion in the offshore environment, so the better solution is to select the gearbox in stainless steel 316.
  B.   The hub shoulder under the clamp should be coated with a strong coating like TSA to prevent corrosion in the offshore environment.
  C.   Subsea valves and actuators are coated with two-component epoxy, but two-component epoxy is not a suitable coating system for an ROV bucket and position indicator.
  D.   Valve coating should be applied after the valve has undergone a pressure test.

  Answer: Option A is correct since it has been experienced frequently that coating a gearbox in steel or cast iron with inorganic zinc is unsuccessful; the coating can be removed from the metal surface and the steel or cast iron gearbox can be corroded easily, as these materials are not corrosion-resistant. The alternative solution is to have the gearbox in stainless steel 316. Option B is not correct, because it is not recommended to coat the hub shoulder under the clamp since the coating can be damaged by the clamp load. Option C is partially correct, because the ROV bucket and position indicator are also coated with two-component epoxy. Option D is correct; the valve coating should be applied after the pressure test. If the body pressure test is performed when the valve is coated, the coating could hide possible leakage from the valve body. Thus, options A and D are completely correct.

6. Fill in the gaps with the correct words.

A mechanical joint includes hubs, clamps, seal ring, bolts and nuts. The uninsulated hubs are in 22Cr duplex, and the operating temperature is 120°C. The applicable coating on the hubs is _____. The clamp is in AISI 4140, and two options are available for coating the clamp; one is _____ and the other is _____. The preferred coating between these two is _____, since it provides higher corrosion resistance. The bolts could be selected in _____ material instead of low-alloy steel bolting with _____ coating to be more robust and corrosion-resistant.

A. Two-component epoxy, TSA, HDG, HDG, duplex, TSA
B. TSA, TSA, HDG, TSA, super duplex, HDG
C. Zinc epoxy, two-component epoxy, TSA, TSA, duplex, HDG
D. Phenolic epoxy, two-component epoxy, HDG, two-component epoxy, super duplex, TSA

Answer: The hub is uninsulated and the operating temperature is higher than 110°C, so the applicable coating is TSA. The clamp is in AISI 4140, low-alloy steel, and can be coated with either TSA or HDG. The preferred coating system is TSA since it provides higher corrosion resistance. The bolts could be selected in super duplex instead of low-alloy steel with HDG as a more robust choice. Thus, option B is the correct answer.

7. Which subsea components in connection with valves and actuators are coated with two-component epoxy?
A. Subsea valves
B. Subsea actuators
C. ROV buckets and position indicators
D. All options are correct

Answer: All four options are correct, so option D is the correct answer.

8. Which sentence is correct regarding cross-contamination?
A. Cross-contamination causes carbon steel corrosion in contact with stainless steel materials.
B. The best way to prevent cross-contamination is to store carbon steel and stainless steel together.
C. Teflon coating powder can be applied on a stainless-steel valve to prevent cross-contamination.
D. Cross-contamination can damage the carbon oxide protective layer on the stainless steel.

Answer: Option A is not correct because cross-contamination causes the corrosion of stainless-steel material in contact with carbon steel.

Option B is not correct, because storing carbon steel with stainless steel results in cross-contamination and corrosion of the stainless steel, so they should be stored separately. Option C is correct; Teflon coating powder can be applied on stainless steel valves to prevent cross-contamination. Option D is not correct, because cross-contamination damages the chromium oxide protective layer on stainless steel. In fact, there is no carbon oxide layer on stainless steel. Therefore, option C is the correct answer.

9. Which sentence is correct regarding valve and actuator coating for fire protection?

   A. The installation of fire nuts for actuators is beneficial since there is no need to apply an insulation box around actuators or to apply coating.

   B. Single-acting actuators may require PFP to be operable in case of fire.

   C. PFP is always applied as a type of coating, e.g., NORSOK coating system 5.

   D. The handwheels and gearboxes of valves in firewater systems are typically coated in red for the purpose of color coding.

   Answer: Option A is not correct because fire nuts are installed around the bolts and nuts of valves. In fact, fire nuts are not applicable for actuators. However, the installation of fire nuts around the valve bolting has many advantages. The main advantage is that there is no need to place a heavy and bulky firebox around the valve. In addition, there is no need to apply coating on the valve under the firebox to prevent CUI. Option B is not correct, because single-acting actuators have springs so, in case of fire, they return to their fail-safe position. Thus, PFP is more applicable for double-acting actuators in offshore, since they work with hydraulic oil and have no spring. Thus, double-acting actuators cannot function properly in the event of fire. Option C is not correct, because PFP could be in the form of a box or insulation and not necessarily a coating system. Option D is correct because it is common to use red color at least on the handwheels and gearboxes of valves in fire water systems for the purpose of color coding.

10. Which types of external corrosion protection are not a type of paint?

    A. PTFE powder coating

    B. PFP insulation

    C. HDG

    D. Zinc-rich primer

    Answer: Option A, PTFE power coating, is not a kind of paint but it can be used on stainless steel materials to prevent the cross-contamination of stainless steel in contact with carbon steel. Option B, PFP, is not a type of paint and is instead a kind of physical barrier installed around components such as valves and actuators to protect them in case of fire. HDG is not considered a type of paint; HDG can be coated on carbon steel

and alloy steel piping components such as bolts, nuts and clamps to protect them against external corrosion. However, HDG is not considered a
strong corrosion protection compared to coating or painting systems like
zinc-rich coating, phenolic epoxy or TSA. Option D, zinc-rich primer, is
a type of coating or paint. Thus, options A, B and C are not considered
a type of coating or paint.

BIBLIOGRAPHY

1. American Petroleum Institute (API) 6D (2012). *Standard for Actuator Sizing and Mounting Kits for Pipeline Valves*, 1st edition. API, Washington DC.
2. American Petroleum Institute (API) 6DX (2020). *Standard for Actuators and Mounting Kits for Valves*, 2nd edition. API, Washington DC.
3. American Petroleum Institute (API) 20E (2012). *Alloy and Carbon Steel Bolting for Use in the Petroleum and Natural Gas Industries*, 1st edition. API, Washington, DC.
4. American Petroleum Institute (API) 20F (2015). *Corrosion Resistant Bolting for Use in the Petroleum and Natural Gas Industries*, 1st edition. API, Washington, DC.
5. American Society for Testing and Materials (ASTM) F467 (2018). Standard specification for nonferrous nuts for general use. West Conshohocken, PA.
6. American Society for Testing and Materials (ASTM) F468 (2016). Standard specification for nonferrous bolts, hex cap screws, socket head cap screws, and studs for general use. West Conshohocken, PA.
7. NORSOK M-001 (2014). Material Selection, 5th edition. Lysaker, Norway.
8. NORSOK M-501 (2012). Surface Preparation and Protective Coating, 6th edition. Lysaker, Norway.
9. NORSOK L-001 (2017). Piping and Valves, 4th revision. Lysaker, Norway.
10. Sotoodeh, K. (2018). Valve failures, analysis and solutions. *Valve World Magazine*, Vol. 23, No. 11, pp. 48–52.
11. Sotoodeh, K. (2018). Standardization for coatings: The KISS-principle. *Valve World Magazine*, Vol. 23, No. 3, pp. 60–63.
12. Sotoodeh, K. (2019). NAB versus titanium seawater system valves. *Valve World Magazine*, Vol. 24, No. 10, pp. 1–3.
13. Sotoodeh, K. (2021). *A Practical Guide to Piping and Valves for the Oil and Gas Industry*. Elsevier (Gulf Professional Publishing), Austin.
14. Sotoodeh, K. (2021). *Subsea Valves and Actuators for the Oil and Gas Industry*. Elsevier (Gulf Professional Publishing), Austin.

Index

A

abrasion 106, 110, 120, 129, 133
abrasive blasting 72–73, 77–81, 85–87, 97
abrasive grit 85, 126, 129, 136
acid pickling 75–76, 99, 102
actuator 181–195, 204–206, 211–212, 217,
　　222, 224, 226–237, 240–241,
　　243, 258–260, 267–273, 276–277,
　　282–288
　electrical 230–231, 233–234, 236, 240–241,
　　258–259, 267–271
　electro-hydraulic 227, 231, 234, 241
　gas-over-oil 227, 231, 235, 240
　hydraulic 191, 229–230, 234–235, 236, 241,
　　268, 271–272
　pneumatic 194, 227, 240–241
additives 106, 110, 141
air 52–53, 57, 74, 79, 81, 84, 87, 91, 195, 206,
　　227–229, 231–233, 236
air compressor 79
aluminum silicate 117
American Society for Testing and Materials
　　(ASTM) 37, 99, 113, 169–170,
　　173, 178
amine 32
American Petroleum Institute (API) 37, 47,
　　202–203, 211, 241
American Society of Mechanical Engineers
　　(ASME) 15, 48, 79, 100, 206,
　　241, 254
anode 5–8, 24, 26, 37–39, 44–46, 55, 76,
　　108, 116
　sacrificial 37, 55
anti-fouling 111–112, 130
aluminum silicon 123, 249
atom 5, 39–40, 76, 121
axial check valve 47, 219–221, 223, 239
axial control valve 213, 215, 224, 238
axial nozzle check valve 266–267, 280
axial valve 215, 219, 223, 225, 239

B

backflow 181, 217, 219, 221, 225
ball 191–199, 206, 210, 269
ball valve 21, 24, 191–199, 203–204, 206,
　　209–211, 223–224, 227, 229,
　　233–240, 248, 250, 257–269,
　　271–274, 282, 284

barrier protection 103–105, 116
bidirectional 213, 263–264
bimetallic corrosion 24
binder 103, 106, 108–110, 116, 119, 131, 134,
　　141–142
　chemically drying 106
　oxidatively drying 106
　physically drying 106
biofouling 12, 112
biological fouling 4, 112
blast cabinet 81–82
blast cleaning 84–85, 87–89, 92–100, 102, 113,
　　15, 153, 168, 178
　brush-off blast cleaning 89, 95, 100
　commercial blast cleaning 87, 95
　industrial blast cleaning 89, 95
　near-white metal blast cleaning 89, 95
　white metal blast cleaning 75, 89, 95
blasting 52, 55, 57, 59, 70–89, 91–92, 96–100,
　　113, 115, 126, 128–129, 132–133,
　　137, 145, 148, 155, 159, 161–162,
　　168, 178, 281
　dry abrasive blasting 72, 81
　light blasting 85
　sand blasting 52, 55, 70, 73–84, 91, 93,
　　96–100, 113, 115, 129, 131–133, 145,
　　148, 154–155, 159, 161–162, 168,
　　178, 281
　shot blasting 77, 79, 100
　sweep blasting 85–87, 89, 100, 126, 128
blast pot 81–82, 84
blast room 81–82
bleeding 155–156, 174–175
blistering 58, 96–97, 145–146, 153, 155,
　　174–176
bluing 55
blush 147, 174
blushing 147, 174
body 11, 23, 127–128, 130, 191–192, 198–199,
　　201, 204–218, 222, 245, 247–249,
　　257–259, 262, 264–266, 276, 278,
　　284–286
bolt 8–9, 12–13, 17–18, 33–35, 37, 46–47,
　　54–55, 127–128, 192, 197–199,
　　204, 211–212, 248–252, 254–258,
　　274–276, 284–288
　plug 253–254
　tensioning 33
bolting specification level (BSL) 275

289

Printed in the United States
by Baker & Taylor Publisher Services